*Jumpstart CMM®/CMMI®
Software Process Improvement*
Using IEEE Software Engineering Standards

Jumpstart CMM®/CMMI® Software Process Improvement
Using IEEE Software Engineering Standards

Susan K. Land

A JOHN WILEY & SONS, INC., PUBLICATION

Copyright © 2005 by IEEE Computer Society. All rights reserved.

Published by John Wiley & Sons, Inc., Hoboken, New Jersey.
Published simultaneously in Canada.

No part of this publication may be reproduced, stored in a retrieval system or transmitted in any form or by any means, electronic, mechanical, photocopying, recording, scanning or otherwise, except as permitted under Section 107 or 108 of the 1976 United States Copyright Act, without either the prior written permission of the Publisher, or authorization through payment of the appropriate per-copy fee to the Copyright Clearance Center, Inc., 222 Rosewood Drive, Danvers, MA 01923, (978) 750-8400, fax (978) 646-8600, or on the web at www.copyright.com. Requests to the Publisher for permission should be addressed to the Permissions Department, John Wiley & Sons, Inc., 111 River Street, Hoboken, NJ 07030, (201) 748-6011, fax (201) 748-6008.

Limit of Liability/Disclaimer of Warranty: While the publisher and author have used their best efforts in preparing this book, they make no representation or warranties with respect to the accuracy or completeness of the contents of this book and specifically disclaim any implied warranties of merchantability or fitness for a particular purpose. No warranty may be created or extended by sales representatives or written sales materials. The advice and strategies contained herein may not be suitable for your situation. You should consult with a professional where appropriate. Neither the publisher nor author shall be liable for any loss of profit or any other commercial damages, including but not limited to special, incidental, consequential, or other damages.

For general information on our other products and services please contact our Customer Care Department within the U.S. at 877-762-2974, outside the U.S. at 317-572-3993 or fax 317-572-4002.

Wiley also publishes its books in a variety of electronic formats. Some content that appears in print, however, may not be available in electronic format.

Library of Congress Cataloging-in-Publication Data is available.

ISBN 0-471-70925-5

Printed in the United States of America.

10 9 8 7 6 5 4 3 2 1

Contents

Preface	xi
Acknowledgments	xiii
Tables and Figures	xv

1 Introduction and Overview 1
   Introduction 1
      What Are the CMM® and CMMI®? 2
      What the CMM® and CMMI® Are Not 2
      What Are Standards? 3
   IEEE Software Engineering Standards 3
      Motivation for IEEE Standards 3
      Categories of IEEE Standards 4
      IEEE Standards Development 5

2 Summary of SW-CMM® 7
   The CMM® for Software (SW-CMM®) 7
      Structural Elements of the SW-CMM® V 1.1 SW-CMM® Maturity Levels ... 7
   Key Process Areas 7
      Common Features 8
      Overview of SW-CMM® Level 2 Key Process Areas 9
      Software Subcontract Management 12
   Appraisal of the CMM® 12
      Capability Maturity Model-Based Assessment Internal Process Improvement ... 12
      CMM® Appraisal Framework 12
      Software Capability Evaluation 13

3 Summary of CMMI-SW® (Staged) — 15

- The CMMI®-SW — 15
  - CMMI®-SW Continuous Versus Staged — 15
  - Structural Elements of the CMMI®-SW (Staged) — 16
- Key Process Areas — 16
  - Process Management — 17
  - Project Management — 17
  - Engineering — 17
  - Support — 18
  - Specific and Generic Goals — 18
  - Specific and Generic Practices — 18
  - CMMI®-SW (Staged) Components — 19
  - Required Components — 19
  - Expected Components — 19
  - Informative Components — 20
  - CMMI®-SW (Staged) Common Features — 20
- Overview of CMMI®-SW (Staged) Level 2 Process Areas — 21
  - Requirements Management — 21
  - Project Planning — 21
  - Project Monitoring and Control — 21
  - Process and Product Quality Assurance — 21
  - Configuration Management — 22
  - Supplier Agreement Management — 22
  - Measurement and Analysis — 22
- Appraisal of the CMMI® — 22
  - ARC — 24
  - SCAMPI — 24

4 Differences between CMM® and CMMI-SW® (Staged) — 27

- SW-CMM® Versus CMMI®-SW (Staged) — 27
  - Brief History — 27
  - The IDEAL Model — 28
- CMM®/CMMI-SW® (Staged) A Maturity Level Comparison — 28
  - Requirements Management — 29
  - Project Planning — 29
  - Project Monitoring and Control — 20
  - Process and Product Quality Assurance — 20
  - Configuration Management — 20
  - Measurement and Analysis — 20
- Why the Move from CMM® to CMMI® — 31

5 IEEE Software Engineering Standards — 33

- Requirements Management — 33
  - The Goals for CMM® Requirements Management — 33
  - The Goals for CMMI®-SW (Staged) Requirements Management — 34
  - Supporting IEEE Software Engineering Standards — 34
    - IEEE Software Requirements Specification IEEE Std 830 — 34

IEEE Software Requirements Specification IEEE Std 1233	34
Requirements Management Analysis	41
Example of IEEE KPA Support for Requirements Management	41
Requirements Traceability	42
Change Enahacement Requests	42
The Goals for CMM® Requirements Management Revisited	45
SW-CCM® Goals for Software Requirements Management	45
CMMI-SW® (Staged) Goal for Requirements Management	46
Software Project Planning	46
The Goals for CMM® Software Project Planning	47
The Goals for CMMI®-SW (Staged) Project Planning	47
Supporting IEEE Software Engineering Standards	48
IEEE Software Project Management Plan IEEE Std 1058	48
IEEE Standard for Software Test Documentation IEEE Std 829	58
IEEE Standard for Software Management IEEE Std 1219	58
Project Planning Analysis	58
Example of IEEE KPA Support for Software Project Planning	58
The Goals of Software Project Planning Revisited	60
SW-CMM® Goals for Software Project Planning	60
CMMI®-SW (Staged) Goals for Project Planning	60
Software Project Tracking and Oversight	60
The Goals of Software Project Tracking and Oversight	61
IEEE and CMM® Software Project Tracking and Oversight	61
IEEE Standard for Software Quality Metrics Methodology IEEE Std 1061™-1998	62
Software Project Tracking and Oversight Analysis	67
Example of IEEE KPA Support for Project Tracking and Oversight	67
The Goals of Software Project Tracking and Oversight Revisited	67
Project Monitoring and Control	68
The Goals of Project Monitoring and Control	68
IEEE and CMMI®-SW (Staged) Software Project Monitoring and Control	68
IEEE Standard for Software Reviews IEEE Std 1028	68
Project Monitoring and Control Analysis	72
Example of IEEE KPA Support for Project Monitoring Control	72
The Goals of Project Monitoring and Control Revisited	72
Software Quality Assurance	73
The Goals for CMM® Software Quality Assurance	73
The Goals for CMMI®-SW (Staged) Process & Product Quality Assurance	74
Supporting IEEE Software Engineering Standards	75
IEEE Standard for Software Quality Assurance Plans IEEE Std 730-2002	75
IEEE Guide for Software Quality Assurance Planning IEEE Std 730.1-1998	75
Software Quality Assurance Analysis	81
Example of IEEE KPA Support for Software Quality Assurance	
The Goals of Software Quality Assurance Revisited	81
SW-CMM® Goals for Software Quality Assurance	81
CMMI®-SW (Staged) Goals for Process and Product Quality Assurance Revisited	83

Software Configuration Management	84
The Goals for CMM® Software Configuration Management	84
The Goals for CMMI®-SW (Staged) Configuration Management	85
Supporting IEEE Software Engineering Standards	86
IEEE Standard for Software Configuration Management Plans IEEE Std 828-1998	86
Software Configuration Management Analysis	92
Example of IEEE KPA Support for Software Configuration Management	93
The Goals of Software Configuration Management Revisited	94
SW-CMM® Goals for Software Configuration Management	94
CMMI®-SW (Staged) Goals for Configuration Management	95
Software Subcontract/Supplier Management	95
The Goals for CMM® Software Subcontractor Management	95
The Goals for CMMI®-SW (Staged) Supplier Agreement Management	96
Supporting IEEE Software Engineering Standards	97
IEEE Recommended Practice for Software Acquisition IEEE Std 1062-1998	97
SW-CMM® Software Subcontractor Management Analysis	105
CMMI®-SW (Staged) Supplier Agreement Management Analysis	105
Example of IEEE KPA Support for Software Subcontractor Management	106
The Goals of Software Subcontractor Management Revisited	107
SW-CMM® Goals for Software Subcontractor Management	107
CMMI®-SW (Staged) Goals for Supplier Agreement Management	107
Measurement and Analysis	108
The Goals of CMMI® Measurement and Analysis	108
Supporting IEEE Software Engineering Standards	109
IEEE Standard for Developing Software Life Cycle Processes IEEE Std 1074™-1997	109
IEEE Standard IEEE Standard Dictionary of Measures to Produce Reliable Software IEEE Std 982.1	109
IEEE Standard Classification for Software Anomalies IEEE Std 1044™-1993 (R2002)	109
IEEE Standard for Software Productivity Metrics IEEE Std 1045-1992 (R2003)	109
Analysis of Measurement and Analysis	113
Proposed Document Outline	114
Example of IEEE KPA Support for Measurement and Analysis	114
The Goals of CMMI® Measurement and Analysis Revisited	115
6 Using IEEE Standards to Achieve Software Process Improvement	**117**
IEEE Supported Process Improvement	117
Define and Train the Process Team (Initiate)	117
Software Engineering	117
SWEBOK	119
Certification	121
Set Realistic Goals (Diagnose)	124
Fix Timelines (Establish)	125
Baseline and Implement Processes (Act)	126

    Perform Gap Analysis (Learn)   127
    Perform self-audit using SW-CMM® KPAs   128
    Perform self-audit using CMMI®-SW (Staged) KPAs.   128
Implementation Pitfalls   129
    Being Overly Prescriptive   129
    Remaining Confined to a Specific Stage   129
    Documentation, Documentation   130
    Lack of Incentives   130
    No Metrics   130
Conclusion   130

Appendix A   IEEE Standards Abstracts   133

Customer and Terminology Standards   133
Life Cycle Standards   136
Process Standards   137
Resource and Technique Standards   139
Product Standards   142

Appendix B   Level 2 Mappings of CMMI-SE/SW/IPPD® (Staged)   145
    V.1.1 to SW-CMM® V. 1.1

References   165

IEEE Publications   165
SEI Publications   168
OTHER References   168

Index   171

About the Author   175

Preface

The idea for this book came from the Institute of Electrical and Electronic Engineers (IEEE) Users of Software Engineering Standards Survey Workshop held at International Software Engineering Standards Symposiums ISESS'97 and ISESS'99. The focus of each of these workshops was to evaluate the findings from two IEEE Standards Users' Surveys, which were conducted over the Internet. An observation that was common to both sets of survey results was that many users indicated that they viewed IEEE standards primarily as reference material to develop their own internal plans. Users indicated that the standards were tailored and used to develop internal documentation for compliance measures. The user community indicated consistently that there was value added in the use of the IEEE software engineering standards set.

This book is intended as a guide for those interested in understanding the value added by the use of selected IEEE Software Engineering Standards when implementing the Capability Maturity Model (CMM) Level 2 software-process improvement methodology.

Acknowledgments

I would like to thank my husband for his unwavering support and encouragement.

I would like to acknowledge my company, Northrop Grumman IT/TASC, for their continued support of my IEEE Computer Society volunteer activities. During this time of corporate cutbacks and budget cuts, their consistent support remains a demonstration of their commitment to excellence through research and standardization.

I would also like to thank my colleagues within the volunteer organizations of the IEEE Computer Society for their open minds, mentoring spirits, and willingness to always find room in the discussion for one more voice.

Tables and Figures

Table 1-1.	What do IEEE Standards do? Some examples.	4
Table 2-1.	Matrix of SW-CMM® KPAs and supporting key practices	10
Table 2-2.	CMM® key process areas by maturity level	11
Table 3-1.	CMMI® Continuous representation capability levels	16
Table 3-2.	CMMI® Staged representation maturity levels	16
Table 3-3.	Matrix of CMMI®-SW (Staged) KPAs and supporting key practices	17
Table 3-4.	Example CMMI® specific practice	19
Table 3-5.	Example CMMI® generic practice	19
Table 3-6.	Examples of CMMI® component categories	20
Table 3-7.	CMM® versus CMMI® common feature comparison	20
Table 3-8.	Comparison of SW-CMM® and CMMI®-SW (Staged) maturity levels	22
Table 3-9.	Summary of the characteristics of the three CMMI® appraisal classes	23
Table 3-10.	SCAMPI modes of usage	24
Table 5-1.	Requirements traceability matrix example	42
Table 5-2.	Typical elements in a CER	43
Table 5-3.	IEEE Std 730 minimum set of software reviews	83
Table 5-4.	Overview of IEEE Std 828-1998	86
Table 5-5.	Cross-reference to IEEE Std 1042-1987	87
Table 5-6.	IEEE Std 1062-1998 for software acquisition—checklists	97
Table 5-7.	IEEE Std 1062 description of software acquisition milestones	105
Table 5-8.	Suggested acquisition plan documentation format	106
Table 5-9.	IEEE Std 1063 Steps for software acquisition	108
Table 5-10.	IEEE Std 982.1 List of measures for reliable software	110
Table 6-1.	10 SWEBOK Areas	120
Table 6-2.	Key software engineering questions	121
Table 6-3.	IEEE standards and training	122
Table 6-4.	The IEEE Certified Software Development Professional (CSDP) exam specifications	123
Table 6-5.	Determine if essential processes are missing or are incomplete	125

Table 6-6.	Examples from SW-CMM® Maturity Questionnaire	126
Table 6-7.	Example implementation timeline	127
Table 6-8.	Example action plan	127
Table 6-9.	SW-CMM® and CMMI® appraisals	128
Table 6-10.	Example of SW-CMM® compliance matrix	128
Table 6-11.	Example of CMMI®-SW (Staged) compliance matrix	129
Table 6-12.	Level 2 CMM®/IEEE standards high-level support matrix	131
Table 6-13.	Level 2 CMMI®-SW (Staged)/IEEE standards comparison matrix	132

Figure 1-1.	IEEE S2ESC standards support of software engineering.	5
Figure 2-1.	The CMM® structure	8
Figure 2-2.	SW-CMM® maturity levels.	9
Figure 3-1.	Overview of the CMMI®-SW (Staged).	18
Figure 4-1.	The IDEAL model	28
Figure 4-2.	Five SW-CMM®/CMMI®-SW (Staged) maturity levels	31
Figure 5-1.	Modified SRS format (IEEE Std 830-1998)	44
Figure 5-2.	Modified SysRS format (IEEE Std 1233-1998)	45
Figure 5-3.	Example SPMP	59
Figure 5-4.	Example SQAP—Based on IEEE Std 730/730.1	82
Figure 5-5.	Example SCM plan	93
Figure 5-6.	Example outline for metrics plan	114
Figure 6-1.	Standards of continuous process improvement	131

1

Introduction and Overview

INTRODUCTION

Many companies, in their push to complete successful Level 2 Capability Maturity Model (CMM®)[1] or Capability Maturity Model Integration (CMMI®)[2] appraisals, have spent large sums of capital to develop and document their software processes. Many times, there is confusion regarding just what each process should contain in order to be defined as one that meets the basic Level 2 criteria as specified in the CMM® or CMMI®. IEEE standards can be used as tools to help with the process definition and documentation required in support of software process improvement. Many of the IEEE software engineering (SE) standards provide detailed procedure explanations, offer section-by-section guidance on building the necessary documentation, and, most importantly, they provide best-practice guidance as defined by those from the software development industry who sit on the panels of standards reviewers.

The CMM® for software (SW-CMM®) and CMMI®-SW are compendiums of software engineering practices that act as motivators for the continuous evolution of improved software engineering processes. It is the premise of this book that IEEE software engineering standards can be used to provide the basic beginning framework for this type of process improvement. IEEE software engineering standards, as a set, can be used to help companies define themselves as Level 2 organizations.

Moving an organization from the chaotic environment of free-form software development toward a more controlled and documented process can be overwhelming to those tasked to make it happen. This book specifically addresses how IEEE standards may be used to facilitate the development of internal plans and procedures in support of repeatable software engineering processes, or SW-CMM®/CMMI®-SW Level 2. It describes

---

[1]CMM® is registered in the U.S. Patent and Trademark Office by Carnegie Mellon University.
[2]CMMI® is registered in the U.S. Patent and Trademark Office by Carnegie Mellon University (August 2002).

how IEEE software engineering standards can be used to help support the definition of best practices.

This book takes the CMM®/CMMI®-SW (Staged) Level 2 process representation and maps it to information supporting goals and practices found in the IEEE standards. The assumption is made that the standards are implemented as is, with no tailoring. This provides the reader with information regarding the value added by using the IEEE standards to implement and define software process. The identification of the strengths and weaknesses of these standards is a by-product of this comparison.

The CMM® and CMMI® do not tell the user "how" to satisfy their KPA criteria. The CMM® and CMMI® are descriptive. They do not describe how to accomplish their goals but describe the criteria that the end results should support. IEEE standards are prescriptive. These standards describe how to full fill the requirements associated with effective software project management.

It is often hard to separate the details associated with software development from the practices required to manage the effort. Simply handing the CMM®/CMMI® to a project leader or manager provides them with a description of an end result. Pairing this with IEEE standards provides them with a way to work toward this desired end. IEEE standards do not offer a "cookie cutter" approach to software management; rather, they support the definition of the management processes in use by describing what is required.

For organizations that do not wish to pursue CMM®/CMMI® Level 2 accreditation, this book will show how the application of IEEE standards, and their use as reference material, can facilitate the development of sound software engineering practices. This book is geared for the CMM®/CMMI® novice, the project manager, and practitioner who wants a one-stop source—a helpful document that provides the details and implementation support required when targeting CMM®/CMMI® implementation with the aid of IEEE software engineering standards.

What Are the CMM® and CMMI®?

The CMMI® (and in a more limited sense, the CMM®) are process frameworks. They:

- Contain the essential elements of effective processes for one or more disciplines
- Contain a framework that provides the ability to generate multiple models and associated training and assessment materials. These models may represent
  - Software and systems engineering
  - Integrated product and process development
  - New disciplines
  - Combinations of disciplines
- Provide guidance to use when developing processes

What the CMM® and CMMI® Are Not

The CMM® and CMMI® models are not processes or process descriptions. Actual processes depend on

- Application domain(s)
- Organization structure

- Organization size
- Organization culture
- Customer requirements or constraints

What are Standards?

Standards are consensus-based documents that codify best practice. Consensus-based standards have seven essential attributes that aid in process engineering. They

- Represent the collected experience of others who have been down the same road
- Tell in detail what it means to perform a certain activity
- Can be attached to or referenced by contracts
- Help to assure that two parties attach the same meaning to an engineering activity
- Increase professional discipline
- Protect the business and the buyer
- Improve the product

IEEE SOFTWARE ENGINEERING STANDARDS

IEEE software engineering standards provide a framework for documenting software engineering activities. The "soft structure" of the standards set lends itself well to the instantiation of CMM® and CMMI®-SW (Staged) Level 2 KPAs. The structure of the IEEE software engineering standards set provides for tailoring. Each standard describes recommended best practices detailing required activities. These standards documents provide a common basis for documenting organizationally unique software process activities.

Motivation for IEEE Standards

When trying to understand exactly what the IEEE software engineering standards collection is, and what this body of work represents, the following statement (taken from the Synopses of Standards section in the *IEEE Standards Collection, Software Engineering,* 1994 Edition [40]) summarizes it best:

> The main motivation behind the creation of these IEEE standards has been to provide recommendations reflecting the state-of-practice in development and maintenance of software. For those who are new to software engineering, these standards are an invaluable source of carefully considered advice, brewed in the caldron of a consensus process of professional discussion and debate. For those who are on the cutting edge of the field, these standards serve as a baseline against which advances can be communicated and evaluated.

IEEE software engineering standards attempt to capture and distill industry best practices. They consolidate existing technology, establishing a firm foundation for introducing newer technology. They increase the professional discipline though the standardization of evolving technologies and methodologies. The application of IEEE software engineering standards helps to ensure a higher quality product. Application of these standards, while keeping the CMM® or CMMI® in mind, helps to ensure that the production of a higher quality product is consistently reproduced. Applying IEEE software engineering standards and the CMM®/CMMI® processes and procedures together can help users define their software development processes while developing software products (see Table 1-1).

Categories of IEEE Standards

As described in the IEEE *Software Engineering Standards Collection* [49], all standards are prescriptive in nature, containing requirements that must be satisfied. These levels of prescription may be used to categorize the IEEE software engineering standards collection:

1. *Terminology* standards provide definitions and unifying concepts for a collection of standards. In many cases, they do not include any explicit requirements, only the implicit demands of applying a uniform terminology.
2. *Collection guides* do not provide requirements—only information. A collection guide surveys a group of related standards and provides advice to users on how suitable standards may be selected for their use.
3. *Principle standards* provide high-level requirements that might be satisfied in a wide variety of ways. They emphasize goals rather than specific means for achieving the goals.
4. *Element standards* are the most familiar form. They contain requirements more detailed than those of principle standards and prescribe a particular approach to achieving the goals prescribed in a principle standard.
5. *Application guides* emphasize recommendations and guidance. They provide advice on how element standards may be implemented in particular situations.
6. *Technique standards* are the most detailed and prescriptive. They generally describe procedures rather than processes. They provide very specific requirements, presumably for those cases in which small deviations might have large consequences.

Table 1-1 What do IEEE Standards do? Some examples

Standard	Function
IEEE 982.1 Measures for Reliable Software	Specifies techniques to develop software faster, cheaper, better.
IEEE 1008 Unit Testing	Describes "best practices."
IEEE 1061 SW Quality Metrics	Provides consensus validity for techniques that cannot be scientifically validated.
IEEE/EIA 12207 SW Life Cycle Processes	Provides a framework for communication between buyer and seller.
IEEE 1028 SW Reviews	Gives succinct, precise names to concepts that are otherwise fuzzy, complex, detailed and multidimensional [78].

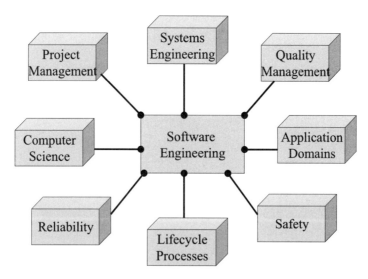

Figure 1-1 IEEE S2ESC standards support of software engineering.

IEEE Standards Development

The focal point for the development and adoption of software engineering standards is the Software and Systems Engineering Standards Committee (S2ESC), a standards sponsor under the IEEE Computer Society Standards Association (IEEE-SA). S2ESC is an organization comprised of over 2000 volunteers who are committed to providing an integrated set of software and systems engineering standards that support the practice of engineering software and systems containing software (Figure 1-1).

The purposes of the S2ESC as defined in the S2ESC Charter[3] are:

1. Codify the norms of professional software engineering practices into standards.
2. Promote use of software engineering standards among clients, practitioners, and educators.
3. Harmonize national and international software engineering standards development.
4. Promote the discipline and professionalization of software engineering.
5. Promote coordination with other IEEE initiatives.

All IEEE software engineering standards are either developed by S2ESC-sponsored working groups or adopted from other standards development organizations. In either case, each standard is submitted to the S2ESC Management Board for a readiness review prior to balloting. Following the readiness review, a balloting pool is formed and the standard is then put forward through a rigorous balloting process. Each negative ballot must be addressed prior to the acceptance of the standard for publication. A cycle of comment resolution, revision, and recirculation continues until consensus is achieved among the balloting group. IEEE defines consensus as 75% approval. Following initial publication, each IEEE standard is subjected to review at least every five years for revision or reaffirmation.

[3]IEEE S2ESC Charter Statement [42].

2

Summary of SW-CMM®

THE CMM® FOR SOFTWARE (SW-CMM®)

In order to understand how IEEE standards fulfill the requirements of the CMM® Level 2 KPAs, it is first important to understand how the CMM® Level 2 maturity level is structured. This structure forms the basis of the matrix comparisons that are presented in the following pages.

Each CMM® maturity level can be decomposed from abstract summaries of each level down to their operational definition in associated key practices (Figure 2-1). Each maturity level is composed of several key process areas. Each key process area is organized into five sections called common features. The common features specify the key practices that, when collectively addressed, accomplish the goals of each key process area.

Structural Elements of the SW-CMM® V 1.1 SW-CMM® Maturity Levels

Increasing levels of software process capability classify the Software CMM® (Figure 2-2). Organizations pursuing process definition and improvement fall within one of five maturity levels. Each maturity level is a step toward achieving a mature software process. Each maturity level indicates a level of process capability. For instance, at Level 2 the process capability of an organization has been elevated from Initial to Repeatable by establishing project management control.

KEY PROCESS AREAS

The SW-CMM® has five maturity levels. Level 2 is the level at which the software process is under basic management control and is fairly repeatable. Each of the maturity

8 SUMMARY OF SW-CMM®

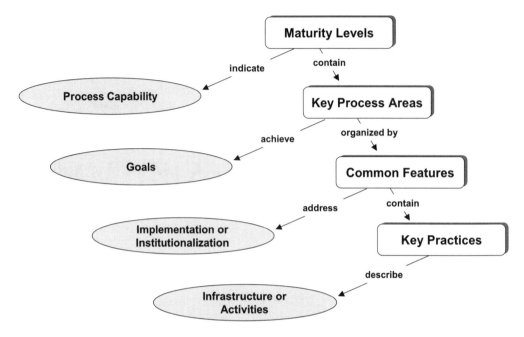

Figure 2-1 The CMM® structure.

levels is composed of collections of goals that are, in turn, supported by key practices. Organizations that have achieved a specific maturity level reach these goals by routinely performing these practices. These collections of software and management practices specific to a maturity level are called key process areas or KPAs.

Common Features

Key practices are structures that are common across the CMM® goals. These key practices are grouped into five categories that are called "common features." Each KPA has all five types of common features. The common features are:

1. *Commitment to perform (Co).* A Commitment to perform practice is usually an organization policy signed by top management.
2. *Ability to perform (Ab).* Ability to perform practices ensure that resources are available to carry out the other practices and that enabling conditions have been satisfied.
3. *Measurement and analysis (Me).* Measurement and analysis practices ensure that the status of the KPA practices is known quantitatively.
4. *Verifying implementation (Ve).* Calls for the regular review by management, or SQA, to ensure that the KPA's implementation is effective and to determine if management intervention is required.
5. *Activities performed (Ac).* Actions the staff might take to carry out the planning, tracking, or training within a process area. Without activities performed, there would be nothing for the other common features to institutionalize.

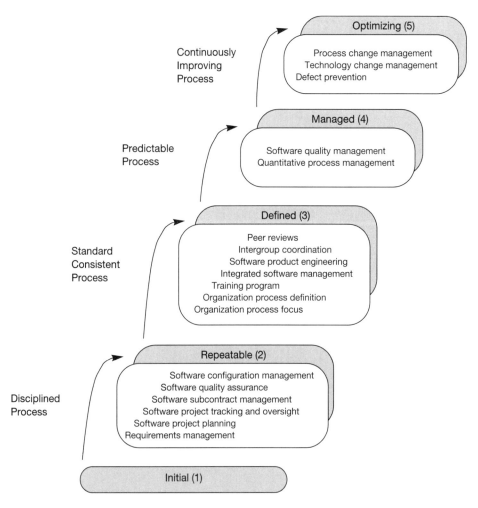

Figure 2-2 SW-CMM® maturity levels [64].

In order for an organization to state with confidence that they are performing at CMM® Level 2, each one of the key practices shown above must be fully met.

There are 18 KPAs distributed over maturity Levels 2–5; there are no KPAs at the initial Level 1. Table 2-1 is a representation of each maturity level and its associated KPAs. Each KPA is composed of key practices whose performance indicates that the KPA is implemented in an organization. To reach Level 2 means that an organization must demonstrate 121 key practices!

Overview of SW-CMM® Level 2 Key Process Areas

The Level 2 KPAs focus on establishing basic project management controls. The challenge is to move from an ad hoc, chaotic, development process to one that is controlled and defined. Organizations that meet these Level 2 criteria are said to have "repeatable" processes. At the repeatable level, organizations demonstrate that their processes are doc-

Table 2-1 Matrix of SW-CMM® KPAs and supporting key practices

Maturity Level	Key Process Area (KPA) Goal	# of Key Practices
5. Optimizing	Defect prevention	18
	Technology change management	19
	Process change management	19
4. Managed	Quantitative process management	18
	Software quality management	13
3. Defined	Organization process focus	16
	Organization process definition	11
	Training program	16
	Integrated software management	19
	Software product engineering	20
	Intergroup coordination	17
	Peer reviews	9
2. Repeatable	Requirements management	12
	Software project planning	25
	Software project tracking and oversight	24
	Software quality assurance	17
	Software configuration management	21
	Software subcontract management	22

umented, enforced, practiced, and measured. They must show that all participants are trained and demonstrate the ability to improve based upon past performance.

All those practicing as software engineers should desire to evolve out of the chaotic activities and heroic efforts of a Level 1 organization. Good software can be produced by a Level 1 organization, but often at the expense of the developers. At the repeatable level, Level 2, software engineering processes are under basic management control and there is a management discipline (Table 2-2).

Requirements Management

The CMM® describes requirements management as the agreement with the customer that forms the basis for planning and managing the software project. Requirements management is that common understanding between the customer and the project team of what is to be accomplished.

Software Project Planning

This KPA describes the type of planning required in support of effective software project management. These plans must be realistic and support project goals and project requirements. An effective plan will form the basis for project software tracking and oversight activities.

Software Quality Assurance

Software quality assurance (SQA) can be defined as the planned and systematic approach to the evaluation of the adherence to software processes and procedures. SQA includes

Table 2-2 CMM® key process areas by maturity level

Processes Categories	Management	Organizational	Engineering
	Software project planning, project management	Senior management review	Requirements analysis, design, develop, test
Levels			
5. Optimizing		Technology change management	
		Process change management	Defect prevention
4. Managed	Quantitative process management		Software quality management
3. Defined	Integrated software Management	Organization process focus	Software product engineering
	Intergroup coordination	Organization process definition	Peer reviews
		Training program	
2. Repeatable	Requirements management		
	Software project planning		
	Software project tracking and oversight		
	Software quality assurance		
	Software configuration management		
	Software subcontract management		
1. Initial	Ad hoc processes		

the process of assuring that standards and procedures are established and are followed throughout the software life cycle. Compliance is evaluated through process monitoring, product evaluation, and audits. The CMM® emphasizes providing management with the appropriate visibility into the supporting process.

Software Configuration Management

This KPA focuses on the identification and documentation of the functional and physical characteristics of a work product. Organizations are required to record and report change processes, baseline status, and change control, and verify compliance against specified requirements.

Software Project Tracking and Oversight

The purpose of software project tracking and oversight is to ensure that project management and senior management have adequate visibility into the software project and its progress. Processes must support the communication to management of any significant deviation from schedule or change in status of project risk. This is to ensure that management is able to take action as soon as possible.

Software Subcontract Management

The focus of this KPA combines the other five KPAs—requirements management, software project planning, software quality assurance, software configuration management, and software project tracking and oversight—and applies them to subcontractor control. An organization is only as strong as its weakest link. The performance of an organization performing at Level 2 will be seriously degraded by the adoption of a subcontractor operating in an ad hoc fashion. The emphasis here is to place the same Level 2 process expectations on any subcontracted organization, requiring the appropriate oversight activities by the hiring entity.

APPRAISAL OF THE CMM®

During a CMM® Level 2 appraisal (assessment or evaluation[4]), individuals attempt to identify the strengths and weaknesses of a software project and its supporting organizational policies. If the evaluation team comes to a consensus that the project and organization support the criteria set forth in the six Key Process Areas (KPAs) for this level, then the project is deemed to be operating at Level 2.

The SEI provides two methodologies to determine the current capabilities of specific organizations: internal assessments and external evaluations. The CBA-IPI (Capability Maturity Model-Based Assessment Internal Process Improvement) was developed to help with self-assessment.

Capability Maturity Model-Based Assessment Internal Process Improvement

The Capability Maturity Model-Based Assessment Internal Process Improvement (CBA IPI) uses the CMM® as a reference model and identifies key process areas for improvement. The CBA IPI is a way for organizations to obtain an accurate picture of the strengths and weaknesses of their current software processes and to validate an organization's commitment to software process improvement. Assessments are an integral part of the diagnosing phase of the IDEAL approach to software process improvement.

CMM® Appraisal Framework

The CMM® Appraisal Framework (CAF) was developed to provide a structure for more formal evaluations. The CAF describes the requirements and guidelines to be used by method developers in designing evaluation methods that are CAF compliant. Just as the CMM® is an abstract model that needs to be instantiated in real practice in a software organization to have meaning, the CAF needs to be instantiated in a real evaluation method. The SEI provides this evaluation methodology in the form of the Software Capability Evaluation (SCE).

---

[4] The term "appraisal" as used at the SEI includes both assessments and evaluations. Both of these focus on an organization's software development process.

Software Capability Evaluation

Software Capability Evaluation (SCE) is used in support of software acquisition as a discriminator to select suppliers, for contract monitoring, and for incentives. They can also be used internally in preparation for an external evaluation. The tool provides an acquirer with an answer to the question, "What is the process capability of this potential supplier?" It determines whether an organization "says what it does and does what it says" by evaluating its software process (usually in the form of policy statements) and project practices. The organization's process captures the "say what you do," and project implementations (specific tailoring and interpretations of this process) should demonstrate the "do what you say." [80]

3

Summary of CMMI-SW® (Staged)

THE CMMI®-SW

After the CMM® for Software (CMM®-SW) was finalized in 1991, many organizations ambitiously pursued software process improvement. The higher levels of the CMM® required that organizations institute enterprise-wide change and support. As organizations changed and improved their processes in support of the CMM® for Software, other CMM® process models were developed. The CMM® Integration (CMMI®) is the consolidation of three source models: The Capability Maturity Model for Software (SW-CMM®) v2.0 draft C, the Electronic Industries Alliance Interim Standard (EIA/IS) 731, and the Integrated Product Development Capability Maturity Model (IPD-CMM®) v0.98. When effectively implemented, the CMMI® supports enterprise cultural change, process definition, and improvement.

CMMI®-SW Continuous Versus Staged

The CMMI® provides for multiple models and two representations: continuous or staged. There are four CMMI® models: systems engineering, software engineering, integrated product and process development, and supplier sourcing. When implementing CMMI® process improvement, the choice of a selected representation can depend upon a number of factors. The continuous representation allows for the selection of the order of improvement, enables comparisons across organizations by process area, provides for the migration from Electronic Industries Alliance Interim Standard (EIA/IS) 731, and provides for comparison to the International Organization for Standardization and International Electrotechnical Commission (ISO/IEC) 15504 Standard.

The staged representation more closely resembles the original CMM® providing for sequential, progressive process improvement. The staged representation allows for organizational comparison across maturity levels and provides a single rating summarizing appraisal results, thereby providing an easy migration from the CMM® for software.

Table 3-1 CMMI® Continuous representation capability levels

Capability Level	Continuous Representation Capability Levels
0	Incomplete
1	Performed
2	Managed
3	Defined
4	Quantitatively managed
5	Optimizing

The continuous representation uses capability levels to measure process improvement. There are six capability levels, numbered 0 through 5. Each capability level corresponds to a generic goal and a set of generic and specific processes (see Table 3-1).

Maturity levels, which belong to the staged representation, apply to an organization's overall maturity. There are five maturity levels, numbered 1 through 5. Each maturity level comprises a predefined set of process areas (see Table 3-2).

The main difference between maturity levels and capability levels is the representation they belong to and how they are applied. The obvious choice when migrating from the CMM® to the CMMI® would be the staged representation of the CMMI® for software engineering, the CMMI®-SW.

Structural Elements of the CMMI®-SW (Staged)

The CMMI® has five maturity levels. Level 2 is the level at which the software process is effectively managed. Each of these maturity levels is also supported by specific practices. There are 21 KPAs distributed over maturity Levels 2–5; there are no KPAs at the initial Level 1. Table 3-3 is a representation of each maturity level and its associated KPAs. To reach Level 2 means that an organization must demonstrate 125 supporting practices!

KEY PROCESS AREAS

In addition to defined levels of maturity, the CMMI® also identifies key process areas (KPAs). These KPAs are grouped into the four categories of process management, project management, engineering, and support.

Table 3-2 CMMI® staged representation maturity levels

Maturity Level	Staged Representation Maturity Levels
1	Initial
2	Managed
3	Defined
4	Quantitatively managed
5	Optimizing

Table 3-3 Matrix of CMMI®-SW (Staged) KPAs and supporting practices

Maturity Level	Key Process Area (KPA) Goal	# of Key Practices
5. Optimizing	Organizational innovation and deployment	19
	Causal analysis and resolution	17
4. Quantitatively Managed	Organizational process performance	17
	Quantitative project management	20
3. Defined	Requirements development	20
	Technical solution	21
	Product integration	21
	Verification	20
	Validation	17
	Organizational process focus	19
	Organizational process definition	17
	Organizational training	19
	Integrated project management	20
	Risk management	19
2. Managed	Requirements management	15
	Project planning	24
	Project monitoring and control	20
	Process and product quality assurance	14
	Configuration management	17
	Supplier agreement management	17
	Measurement and analysis	18

Project Management

The project management process areas cover the project management activities related to the planning, monitoring, and control of projects. The project management Level 2 process areas of CMMI® project planning include project monitoring and control and supplier agreement management.

Engineering

The Engineering process areas group development and maintenance activities that are shared across engineering disciplines (e.g., systems engineering and software engineering). The CMMI® KPA requirements management is the Level 2 type of KPA in support of the engineering process group.

Support

The support process areas concentrate activities supporting product development and maintenance. These support process areas address performance processes, or those used in support of other processes. Level 2 processes included in the support grouping are configuration management, process and product quality assurance, and measurement and analysis.

Process Management

The process management process areas focus on the organizational activities related to the definition, planning, deployment, implementation, monitoring, control, appraisal, measurement, and improvement of processes. There is no CMMI® Level 2 KPA that directly supports this; rather, this KPA category plays a key role in Level 3 process implementation.

Specific and Generic Goals

Each maturity level contains process areas, specific goals, specific practices, generic goals, generic practices, typical work products, subpractices, notes, discipline amplifications, generic practice elaborations, and references (see Figure 3-1).

Specific and Generic Practices

In the staged representation, there are specific and generic goals, which are numbered sequentially. Each specific goal has a number beginning with SG. Each generic goal has a number beginning with GG. The practice title is not used for appraisals or rated in any way. However, the practice statement is used for process-improvement and appraisal purposes.

Specific practices begin with SP, followed by a number in the form x.y. The x is the number of the specific goal that the practice is mapped to, and the y is the practice's sequence number. For example, in the requirements management process area, the first

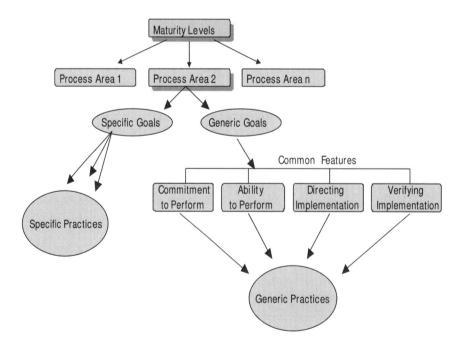

Figure 3-1 Overview of the CMMI®-SW (Staged).

Table 3-4 Example CMMI® specific practice

SG1	**Requirements are managed and incorporated with project plans and work products are identified.**
SP 1.1	*Obtain an Understanding of Requirements; Develop an understanding with the requirements providers on the meaning of the requirements.*

specific practice associated with specific goal 1 is numbered SP 1.1, and the second is SP 1.2 (see Table 3-4).

Generic practices are numbered in a similar way beginning with GP, followed by a number in the form x.y, where x is the number of the generic goal that the practice is mapped to, and y is the practice's sequence number. A second number is used for the generic practices, indicating the sequence number of the practice within one of the four common feature categories to which it belongs. For example, the first generic practice associated with GG 2 is numbered GP 2.1 and CO 1. The CO 1 number indicates that the generic practice is the first generic practice organized under the Commitment to Perform common feature (see Table 3-5).

CMMI®-SW (Staged) Components

The components of a CMMI® model are grouped into three categories: required, expected, and informative (see Table 3-6). All specific goals and generic goals are required model components. These components must be clearly demonstrated by an organization's processes.

Required Components

Required components are essential to rating the achievement of a process area. Specific practices and generic practices are the crucial CMMI® model components. These components clearly define any key process-supporting elements and are fundamental to rating the achievement of a process area. Goal achievement (or satisfaction) is used in appraisals as the basis upon which process area satisfaction and organizational maturity are determined.

Expected Components

Expected components describe what an organization will typically implement to achieve a required component. Expected components guide those implementing improvements or performing appraisals. These components support the required goals for process improvement and are expected to be present in the planned and implemented processes of the organization before goals can be considered satisfied.

Table 3-5 Example CMMI® generic practice

GG2	**Institutionalize a Managed Process**
GP 2.1 (CO1)	*Establish an Organizational Policy; Establish and maintain an organizational policy for planning and performing the requirements management process.*

Table 3-6 Examples of CMMI® component categories

Required	SG 1. Requirements are managed and inconsistencies with project plans and work products are identified.
Expected	SP 1.1. Develop an understanding with the requirements providers on the meaning of the requirements.

Informative	Typical Work Products
	    Lists of criteria for distinguishing appropriate requirements providers; Criteria for evaluation and acceptance of requirements
	    Results of analyses against criteria
	    An agreed-to set of requirements subpractices; criteria established for distinguishing appropriate requirements providers and for the acceptance of requirements.

Informative Components

Informative components provide details that help model users get started in thinking about how to approach goals and practices. These subpractices, typical work products, discipline amplifications, generic practice elaborations, goal and practice titles, goal and practice notes, and references are informative model components that help model users understand the goals and practices and how they can be achieved.

CMMI®-SW (Staged) Common Features

There are four common features used in CMMI® models with a staged representation (see Table 3-7):

1. Commitment to perform (CO) identifies the generic practices relating to the creation of policies and corporate sponsorship.
2. Ability to perform (AB) identifies the generic practices that ensure that a project or organization has adequate designated resources.
3. Directing implementation (DI) defines the group of generic practices relating to the management of process performance, the integrity of associated work products, and stakeholder involvement.
4. Verifying implementation (VE) designates the group of generic practices relating to management review and conformance evaluation.

Table 3-7 CMM® versus CMMI® common feature comparison

CMM®	CMMI®
Commitment to perform (Co)	Commitment to perform (CO)
Ability to perform (Ab)	Ability to perform (AB)
Activities performed (Ac)	
Measurement and analysis (Me)	
Verifying implementation (Ve)	Verifying implementation (VE)
	Directing implementation (DI)

These common features provide a basis for organization for the generic practices of each process area. Common features are model components that provide a way to present the generic practices. Each common feature is designated by an abbreviation as shown in Table 3-7.

OVERVIEW OF CMMI®-SW (STAGED) LEVEL 2 PROCESS AREAS

Requirements Management

The purpose of requirements management is to manage requirements associated with a project and to identify inconsistencies between the requirements and the project plan and associated work products. The CMMI® defines the requirements for the bidirectional traceability of software requirements and specifies that for requirements management to be effective, it must operate in parallel with requirements development, providing adequate support for requirements change. The CMMI®-SW (Staged) also specifically requires that the requirements management process be institutionalized as a managed process.

Project Planning

The CMMI®-SW (Staged) Level 2 is focused on establishing estimates, plan development, and plan commitment in support of project activities. The CMMI® places emphasis on having a detailed project work breakdown structure (WBS). The planning and maintenance of project data items and their contents has also been added to the list of project management concerns. The CMMI® specifically requires the execution of the project according to estimates. Estimation focuses on size and complexity, whereas effort, cost, and schedule are determined and established, respectively, based on these estimates. The CMMI®-SW (Staged) also addresses the reconciliation (revision) of the project plan to reflect changes in available and estimated resources.

Project Monitoring and Control

The focus of this KPA is on the processes supporting the effective management of a software project. This begins with the documentation of the project plan and provides for the required processes supporting the identification and appropriate handling of deviations from the planning documentation. It addresses the effective management of any required corrective action when project performance deviates significantly from projected baselines.

Process and Product Quality Assurance

The Process and product quality assurance process area provides for the definition of the activities associated with software project oversight. This KPA supports all process areas by requiring the objective evaluation of conformance to stated project processes. Process and product quality assurance ensures that the project staff and management have proper visibility into the processes and work products throughout the life cycle of a specified project.

Configuration Management

This KPA focuses on the processes in support of the definition, control, review, and reporting of the work products associated with a software project. It ensures that the integrity of all work products, and all items used to create any work product, is established and effectively maintained. Configuration management (CM) activities ensure that changes to work products are captured and managed. This KPA applies to all other process areas by requiring that all the processes and process artifacts used to support a software effort be placed under configuration management control. CM activities include the management of requirements baselines and supporting project plan revision management source code control.

Supplier Agreement Management

This KPA focuses on the processes supporting the acquisition of products from suppliers with whom there exist formal agreements. The supplier agreement management process area addresses what is required in support of effective acquisition of work produced by suppliers external to an organization. A supplier agreement is established and maintained and is used to manage the supplier. This KPA requires that all progress and performance be monitored and reported.

Measurement and Analysis

The measurement and analysis (MA) KPA defines the processes supporting the development, maintenance, and implementation of software project measurement activities. These activities support objective planning and estimating for all process areas tracking actual performance against initial estimates. The results from MA activities can be used in making informed decisions and taking appropriate corrective actions.

APPRAISAL OF THE CMMI®

For assessments and evaluations, the SEI has published the Standard CMMI® Appraisal Method for Process Improvement (SCAMPI) as a replacement for CBA IPI and SCE, which were used in support of the SW-CMM® (see Table 3-8 for maturity level comparison). Due to the phaseout of the SW-CMM®, the SEI will not produce any updates to the CBA IPI and SCE methods used to support CMM® assessments. The training of Lead Appraisers and SCE Lead Evaluators was supported through December 2003 and all will be required to transition to SCAMPI Lead Appraisers within two years of this date.

Table 3-8 Summary comparison of SW-CMM® and CMMI®-SW (Staged) maturity levels

Maturity Level	SW-CMM®	CMMI®-SW (Staged)
5	Optimizing	Optimizing
4	Managed	Quantitatively managed
3	Defined	Defined
2	Repeatable	Managed

CMMI® assessments and evaluations allow organizations to:

1. Gain insight into their engineering capabilities through the identification of process strengths and weaknesses.
2. Relate process strengths and weaknesses to the CMMI® model
3. Prioritize their improvement plans
4. Focus on improvements that are most beneficial
5. Derive capability and maturity level ratings
6. Identify risk relative to capability/maturity determinations [55]

Table 3-9 Summary of the characteristics of the three CMMI® appraisal classes [64]

Characteristics	Class A	Class B	Class C
Usage mode	• Rigorous and in-depth investigation of process(es) • Basis for improvement activities	• Initial (first-time) • Incremental (partial) • Self-assessment	• Quick-look • Incremental
Advantages	• Thorough coverage • Strengths and weaknesses for each PA investigated • Robustness of method with consistent, repeatable results • Provides objective view	• Organization gains insight into own capability • Provides a starting point or focuses on areas that need most attention • Promotes buy-in	• Inexpensive • Short duration • Rapid feedback
Disadvantages	• Demands significant resources	• Does not emphasize depth of coverage and rigor and cannot be used for level rating	• Provides less buy-in and ownership of results • Not enough depth to fine-tune process improvement plans
Sponsor	• Senior manager of organizational unit	• Any manager sponsoring an SPI program	• Any internal manager
Team composition	• External and internal	• External or internal	• External or internal
Team size	• 4–10 persons + assessment team leader	• 1–6 + assessment team leader	• 1–2 + assessment team leader
Team qualifications	• Experienced	• Moderately experienced	• Moderately experienced
Assessment team leader requirements	• Lead assessor	• Lead assessor or person experienced in method	• Person trained in method

ARC

The Appraisal Requirements for CMMI® (ARC) defines what is required for the development, definition, and use of appraisal methods in support of compliance with the CMMI® [56]. The ARC supports both CMMI® representations, the staged and continuous, and supports assessments (for internal process improvement) and capability evaluations (for source selection and/or process monitoring). The ARC has formalized the requirements for three classes of appraisal methods by mapping CMMI® requirements to each method. It provides organizations the freedom to develop an appraisal methodology that works best for their organization (see Table 3-9).

Class A describes a full appraisal, usually performed by a team of six to 10 people, primarily drawn from inside the organization being appraised. This appraisal method is expected to be the most accurate, will maximize the buy-in from the appraisal participants, and will leave the organization with the best understanding of their weaknesses and any strengths that should be shared. The Standard CMMI® Appraisal Method for Process Improvement (SCAMPI) describes a Class A appraisal [59].

SCAMPI

SCAMPI is designed to provide organizations with quality ratings relative to CMMI® models. It supports both internal process improvement assessments and external capability determinations and satisfies all Class A appraisal ARC requirements. Assessment results can be used to support internal process improvement evaluation, supplier selection, and process monitoring [59] (see Table 3-10).

Table 3-10 SCAMPI modes of usage

Usage Mode	Description	Applications
Internal process improvement	Appraisals are used to evaluate internal processes: • Baseline capability/maturity levels • Establish or update a process improvement program • Measure progress in program implementation	• Measuring progress • Conducting audits • Appraising specific projects • Preparing for external appraisals
Supplier selection	• Discriminator for supplier selection • Baseline for subsequent process monitoring	• Characterize process-related risk of supplier selection • Assist in supplier selection
Process monitoring	• Help the sponsoring organization tailor its contract or process monitoring efforts by allowing it to prioritize efforts based on the observed strengths and weaknesses of the supplying organization's processes • Focus on long-term teaming relationship between sponsor and supplier organizations	• Serving as input for incentive/award fee • Risk management

As an ARC Class A method, SCAMPI can be used to generate ratings as benchmarks to compare maturity levels or capability levels across organizations. SCAMPI is an integrated appraisal method that can be applied in the context of internal process improvement, supplier selection, or process monitoring. SCAMPI offers a rigorous methodology capable of achieving high accuracy and reliability of appraisal results through the collection of objective evidence from multiple sources. It provides detailed guidance on what should be measured and how the assessment should be carried out.

4

Differences Between CMM® and CMMI-SW® (Staged)

SW-CMM® VERSUS CMMI®-SW (STAGED)

Brief History

In November 1986, the Mitre Corporation and the Software Engineering Institute (SEI),[5] under the leadership of Watts Humphrey, were tasked by the U.S. Department of Defense to develop a way to assess the ability of its software contractors. The result of this effort was the Capability Maturity Model for Software (SW-CMM®). Following the deployment of the initial model, the SW-CMM® was quickly embraced by software organizations as a viable way to improve the processes associated with their software development. As a by-product of this adoption, numerous other CMM®s, such as software acquisition (SA-CMM®) were also developed.

Driven by the need to incorporate changes requested by users of the SW-CMM® and the need to integrate the various other CMM® process models, the SEI was again asked to think about process improvement. The U.S. Department of Defense[6] and the Systems Engineering Committee of the National Defense Industrial Association (NDIA) funded the development of the Capability Maturity Model Integration (CMMI®).

The CMM® Integration (CMMI®) Project was conceived as an initiative to combine the various CMM®s into a set of integrated models. The source models that served as the basis for the CMMI® include: CMM® for Software V2.0 (Draft C), EIA-731 Systems Engineering, and IPD CMM® (IPD) V0.98a. The move from the SW-CMM® to the CMMI® requires a shift in focus. The CMM® is written for the software organization embracing a management focus and provides principles for software engineering management. The CMMI® is broader in scope, providing more of an organizational focus.

---

[5]Of Carnegie Mellon University.
[6]Specifically, the Office of the Under Secretary of Defense, Acquisition, Technology, and Logistics (OUSD/AT&L).

The objective of the CMM® and CMMI® developers has been to provide models based upon actual practices reflecting the state of the practice. The CMMI® is an upgrade of the SW-CMM®, adding new process areas, modern best practices, and generic implementation goals that apply to each process area. Both models describe what is required for effective software processes.

The IDEAL Model

The IDEAL (initiating, diagnosing, establishing, acting, and learning) model is an organizational improvement model that was originally developed to support CMM®-based software process improvement. It serves as a roadmap for initiating, planning, and implementing improvement actions. This model can serve to lay the groundwork for a successful improvement effort (initiate), determine where an organization is in reference to where it may want to be (diagnose), plan the specifics of how to reach goals (establish), define a work plan (act), and apply the lessons learned from past experience to improve future efforts (learn) (see Figure 4-1). This model serves as the basis for both the SW-CMM® and CMMI® process improvement methodologies.

CMM®/CMMI -SW® (Staged) A Maturity Level Comparison

When discussing the CMM® and CMMI®, it is important to understand that each of these models are defined by increasing levels of software process maturity. Organizations are measured and their capability determined against these levels of maturity. These levels

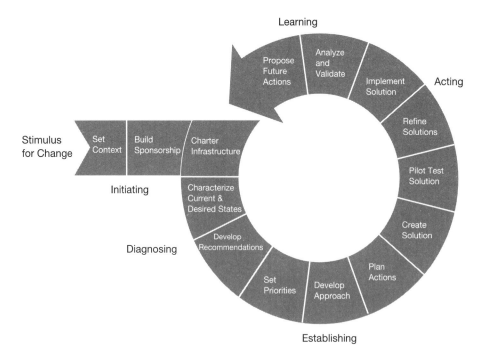

Figure 4-1 The IDEAL model [62].

range from repeatable/managed at Level 2 to optimizing at Level 5, where they demonstrate that their processes are optimized.

For organizations migrating from the CMM® to the CMMI®, or wishing to implement either model, it is important to understand each model, their similarities and differences. IEEE software engineering standards can most effectively be applied, and the most benefit gained by the implementing organization, once the requirements of the CMM®/CMMI® are clearly understood.

Requirements Management

In defining Level 2 software requirements management, the SW-CMM® states that the analysis and allocation of the system requirements is not the responsibility of the software engineering group but is a prerequisite for their work. This process model does not support true integration between system and software engineering nor does it support stakeholder concurrence. The CMMI®-SW (Staged) integrates both systems and software. Integrated teams of all relevant stakeholders are involved in the development of requirements, architecture, and processes concurrently.

The CMMI® explicitly calls for the bidirectional traceability of software requirements. This traceability was addressed in the CMM® but its importance is emphasized in the CMMI®. The CMMI® specifics that for requirements management to be effective, it must operate in parallel with requirements development and offer support as requests for new requirements and requirements change requests are made. A requirement is traceable if its rationale and source are known, all other requirements that are related to it are identified, and its association to other information, such as design, are documented. The CMMI®-SW (Staged) also specifically requires that the requirements management process be institutionalized as a managed process; this is only implied by the SW-CMM®.

Project Planning

The CMMI®, as opposed to the SW-CMM®, emphasizes having a detailed project work breakdown structure (WBS), including a focus on the required supporting knowledge and skills of selected project personnel. The planning and maintenance of project data items and their contents has also been added to the list of project management concerns. The CMMI® specifically requires the execution of the project according to estimation. Estimation focuses on size and complexity, whereas effort, cost, and schedule are determined and established, respectively, based on this estimate. The CMMI®-SW (Staged) speaks to the reconciliation (revision) of the project plan to reflect changes in available and estimated resources, which is not directly address by the SW-CMM®.

The identification and involvement of stakeholders is an important evolution of the "all affected groups" statement that appeared frequently in the CMM® for software. The CMMI® required plan for stakeholder interaction requires a description of all relevant stakeholders, their relationship, the rationale for their involvement, their expected roles and responsibilities, their relative importance to project success by phase, any required supporting resources, and identification in the project schedule. The CMMI®-SW (Staged) requires that there be documentation describing the project planning process. This description of the planning process is not consistently addressed in the SW-CMM®.

Project Monitoring and Control

The CMMI® has elevated the monitoring of project commitments. Monitoring risks and stakeholder involvement is also more strongly emphasized in the CMMI® than in the CMM®. This CMMI® KPA integrates project management and measurement activities. The project management plan becomes the basis for monitoring activities, communicating status, and taking corrective action. Progress is measured against this plan by comparing actual work product and task attributes, effort, cost, and schedule at prescribed milestones within the project schedule or work breakdown structure.

The requirements to manage corrective action to closure when project performance deviates significantly from the project baseline are addressed more rigorously in the SW-CMM®. However, the CMMI®-SW (Staged) specifically requires that the process be institutionalized as a managed process, which is only implied by SW-CMM®. It is also important to note that the SW-CMM® does not directly address the CMMI® requirement for having a plan for performing project monitoring and control activities.

Process and Product Quality Assurance

The CMMI® requires the objective evaluation of products, as well as their supporting processes. It requires that evaluation criteria must be established based upon business objectives identifying what will be evaluated, how often and when the evaluation will occur, how the evaluation will be conducted, documentation of evaluation results, and all key players. The CMMI® places an emphasis on embedded, objective quality assurance activities, ensuring that those performing quality assurance activities are trained and that those performing quality assurance activities for a work product are separate from those directly involved in the development or maintenance of the work product. The CMMI® fully supports the requirement that the processes in support of quality assurance (QA) activities be planned and that the plan(s) documenting the QA processes must be maintained.

Configuration Management

The CMMI® has moved away from the limiting concept of configuration management (CM) as simply tracking software in a library, as described in the SW-CMM®. Configuration management is applied to all items in support of the development effort, including the processes themselves. The CMMI® requires that the concept of a CM system be incorporated into process improvement efforts. The SW-CMM® does not fully support the requirement to establish and maintain the plan for performing the CM process as described by the CMMI®-SW (Staged).

Measurement and Analysis

In the SW-CMM®, measurement and analysis practices were scattered throughout the model, but in the CMMI® they are formalized into the Measurement and Analysis (MA) process area. The CMMI® requires measurement initiatives that specify the objectives of the MA to be performed. The CMMI® requires that an organizational policy for planning and performing the MA process be established. The CMMI® requires that measurements must be planned and aligned with established information needs and business objectives, that adequate resources are provided in support of the MA process, and that responsibility and authority for process performance is clearly defined. The SW-CMM® requirements for establishing and maintaining measurement objectives, which are derived from identified information needs and objectives, are significantly less rigorous.

All measures to be used, data collection processes, associated storage mechanisms, analysis processes, reporting processes, and feedback mechanisms must be defined. As previously stated, the analysis of these measurements should support business decisions and be the motivator for any required corrective action. Correctly implemented, the CMMI® sets an organization on an evolutionary path from basic project management measures, to those based on the organization's set of standard processes, to statistical control of selected subprocesses that support their business needs.

WHY THE MOVE FROM CMM® TO CMMI®

The initial Capability Maturity Model (SW-CMM® v1.0) was developed by the Software Engineering Institute and specifically addressed software process maturity. It was first released in 1990, and after its successful adoption and use, other CMM®s were developed to support process improvement for other disciplines such as systems engineering, people, integrated product development, and software acquisition. Although many organizations found these models to be useful, they also struggled with problems caused by model overlap, inconsistency, and integration that were a result of the many disparate CMM® models. Many organizations also confronted conflicting demands between these models and other process improvement programs, such as ISO 9000 audits.

Both models, the SW-CMM® and the CMMI®, have been embraced by the software industry as a way to define, improve, and demonstrate process capabilities (see Figure 4-2). However, it is important to note that the SW-CMM® has been phased out. Following the publication of CMMI® in December 2001, the SEI has not produced any updates to the SW-CMM® model and training and CMM® assessments are not supported. The CMMI®-SW (Staged) is identified as the closest replacement CMMI® for the SW-CMM®.

Many companies have not made the transition to the CMMI®. This is partially due to

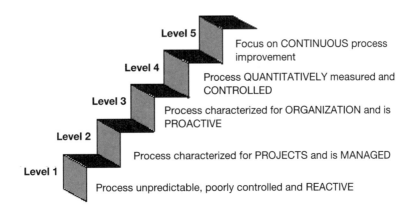

Figure 4-2 Five SW-CMM®/CMMI®-SW (Staged) maturity levels.

the fact that organizations are resistant to change. Many companies have not felt the urgency to move to the CMMI® because the SW-CMM® was supported as a legacy product until the end of 2003. The data from legacy SEI-authorized assessments against the SW-CMM® was still being accepted by the SEI during this period. Both models, the SW-CMM® and the CMMI®-SW (Staged), are presented in this book to illustrate how IEEE standards can benefit both user communities—legacy SW-CMM® and the new CMMI-SW®.

Regardless of which SEI software process model is applied, the SW-CMM® or the CMMI-SW® (Staged), software process maturity may be defined as:

> ... the extent to which a specific process is explicitly defined, managed, measured, controlled, and effective. Maturity implies a potential for growth in capability and indicates both the richness of an organization's software process and the consistency with which it is applied in projects throughout the organization. The software process is well understood throughout a mature organization, usually through documentation and training, and the process is continually being monitored and improved by its users. The capability of a mature software process is known. Software process maturity implies that the productivity and quality resulting from an organization's software process can be improved over time through consistent gains in the discipline achieved by using its software process.[7]

Organizations that have already achieved a high level of maturity may wish to make the transition from the CMM® to the CMMI® more quickly to take advantage of the increased organizational coverage described in CMMI® models. There is significant improvement in coverage of the engineering, risk management, and measurement and analysis processes, as compared to the Software SW-CMM®, and organizations will find strong commonality between CMMI® models and the SW-CMM®.

---

[7]SEI CMM V1.1, p. 4.

5

IEEE Software Engineering Standards

REQUIREMENTS MANAGEMENT

Now we will look at the IEEE standards that support each of the seven process areas of the CMMI® staged representation. We begin with requirements management. Requirements are the basic building blocks of the entire software development process. They are what are to be implemented in the software, they form the basis for the schedule estimates, and they provide a template for acceptance and testing. The CMM® says, quite simply, that the software engineering group uses the software requirements as the basis for plans, products, and activities. The CMM® does not offer much detail, which is where IEEE Standard 830, Requirements Management, comes in.

There are two major activities during the requirements phase of any software effort: problem understanding and requirement specification. Once the essentials of the problem are understood, the requirements must be documented. The requirements document must specify all functional and performance requirements; the formats of inputs and outputs; and all design constraints that may exist. All of the factors that may effect the design and proper functioning of the system should be addressed, not just the required functionality.

The CMM® states that the purpose of requirements management is to establish a common understanding between the customer and the software projects of the customer's requirements that will be addressed by the software project. This agreement is the basis for planning and managing the software project.

The Goals for CMM® Requirements Management

This CMM® KPA emphasizes the management of defined requirements throughout the entire software lifecycle. When system requirements are changed within a Level 2 organization, any affected work plan, product, or activity is adjusted to remain consistent with the change in the requirements.

Effective requirements management establishes an understanding between the customer and the software project of the customer's requirements. The customer may be internal or external to the developing organization.

The customer/producer agreement regarding the software requirements is referred to as the "system requirements allocated to the software" and covers both technical and nontechnical (e.g., schedule) requirements. This allocation of the software requirements forms the basis for all software project activities.

There are two goals for CMM® requirements management:

Goal 1. System requirements allocated to software are controlled to establish a baseline for software engineering and management use.

Goal 2. Software plans, products, and activities are kept consistent with the system requirements allocated to software.

These goals require that all parties involved review the allocated requirements. All software plans, work products, and activities must be revised when the requirements change. Also, changes to any commitments resulting from changes to the allocated requirements must be negotiated.

The Goals for CMMI®-SW (Staged) Requirements Management

The goals in support of the CMMI®-SW (Staged) do not differ greatly from those supporting the CMM® Level 2 requirements. As with the CMM®, the CMMI® requires that the software requirements be defined, agreed upon, and managed throughout the development process. There are two goals:

SG1. Manage Requirements. Requirements are managed and incorporated with project plans and work products are identified.

GG2. Institutionalize a Managed Process. The process is institutionalized as a managed process.

The CMMI® places increased emphasis throughout the model on process institutionalization. Making requirements management part of the corporate culture helps to ensure that the processes in place are not abandoned during crisis situations.

Supporting IEEE Software Engineering Standards

IEEE Software Requirements Specification IEEE Std 830. IEEE Std 830-1998, IEEE Recommended Practice for Software Requirements Specifications, outlines the requirements for what comprises a good software requirements specification (SRS). This document describes in detail the processes supporting the clear and consistent documentation of the software requirements associated with a specific software effort.

There are several advantages to be gained by the use of the IEEE Recommended Practice for Software Requirements Specifications (SRS). First, it is a standard drawn up by representatives of many organizations involved in the elicitation of software requirements. Input is received from industry and university representatives who are the members of the working group and reviewing teams. This collection of individuals represents many years of SRS and development experience. Second, the IEEE SRS is designed to result in an unambiguous software requirements specification process and a complete specification document. Third, the plan provides information on the elicitation, documentation, and communication of the requirements. It acts as a framework for organizations to follow. Each SRS fits a certain pattern, and this makes it easier for others to understand.

IEEE System Requirements Specification IEEE Std 1233. The IEEE Guide to Developing System Requirements Specifications is designed to result in a complete system requirements (SysRS) document. This plan provides information on the elicitation, documentation, and communication of system requirements including an overview of the system requirements specification process. This standard describes what comprises a well-defined system requirement, including a list of pitfalls to avoid when building a set of system requirements.

The following matrix provides a cross-reference of the SEI CMM® Level 2 Requirements Management KPA to relevant supporting documentation from the IEEE Software Engineering Standards Collection. This is not meant to be an exhaustive list, but rather is used to illustrate how the various KPAs are directly supported by IEEE standards.

IEEE/CMM® Requirements Management Matrix

CMM® Goals and Key Practices	Definiton	IEEE Source	Comments
G1	System requirements allocated to software are controlled to establish a baseline for software engineering and management use.		
AC1	The software engineering group reviews the allocated requirements before they are incorporated into the software project.	IEEE Std 1058-1998	Section 4.5.3.1, requirements control plan.
		IEEE Std 830-1998	Recommended practice for software specifications development, 4.4.
		IEEE Std 1233-1998	Guide for systems specifications development.
		IEEE Std 1028-1997 (R2002)	Section 4 and Section 5 address management and technical review of requirements, respectively (Software Reviews and Audits).
		IEEE Std 1074-1997	Section 5, addresses the requirements process; Annex A.2 and A.3.
G2	Software plans, products, and activities are kept consistent with the system requirements allocated to software.		

IEEE/CMM® Requirements Management Matrix (*continued*)

CMM® Goals and Key Practices	Definiton	IEEE Source	Comments
AC2	The software engineering group uses the allocated requirements as the basis for software plans, work products, and activities.	IEEE Std 830-1998	Recommended practice for software specifications development, 4.1.
		IEEE Std 1012-1998	Section 7.5, establish specific milestones for initiating and completing each task, for the receipt of each input, and for the delivery of each output.
		ISO/IEC 12207.1-1996	Sections 6.22, software requirements description.
		IEEE Std 1074-1997	Annex A, A.1.1.3, allocate project resources.
		IEEE Std 830-1998	Section 4.8, embedding project requirements in the SRS.
AC3	Changes to the allocated requirements are reviewed and incorporated into the software project.	IEEE Std 830-1998	Recommended practice for software specifications development, 4.
		IEEE Std 828-1998	Section 4.
		ANSI/IEEE Std 1042-1987 (R1993)	Provides examples and templates.
		IEEE Std 830-1998	Section 4.4, joint preparation of the SRS.
		IEEE Std 1058-1998	Section 4.5.3.1, requirements control plan.
CO1	The project follows a written organizational policy for managing the system requirements allocated to software.	IEEE Std 1028-1997	Provides entry and exit criteria for all procedures for all phases of software life cycle.
			No requirement in standard for policy statement.
AB1	For each project, responsibility is established for analyzing the system requirements and allocating them to hardware, software, and other system components.	IEEE Std 1058-1998	Section 4.5.3.1, requirements control plan.
AB2	The allocated requirements are documented.	IEEE Std 1058-1998	Section 4.5.3.1, requirements control plan.
		IEEE Std 830-1998 and IEEE Std 1233-1998	How to document requirements.

IEEE/CMM® Requirements Management Matrix (*continued*)

CMM® Goals and Key Practices	Definiton	IEEE Source	Comments
AB2 (*cont.*)		IEEE Std 1219-1998	Maintenance requirements.
		IEEE Std 1220-1998	Section 6.1, requirements analysis.
		ISO/IEC 12207.1-1996	Sections 6.22, software requirements description.
AB3	Adequate resources and funding are provided for managing the allocated requirements.	IEEE Std 1058-1998	Section 4.5. This section calls out the requirement to allocate resources and funding for activities, but does not reference requirements management specifically.
AB4	Members of the software engineering group and other software-related groups are trained to perform their requirements management activities.	IEEE Std 1058-1998	Section 4.5.1.4, project staff training plan.
		IEEE Std 1219-1998	Software maintenance activities. Section 4.
ME1	Measurements are made and used to determine the status of the activities for managing the allocated requirements.	IEEE Std 830-1998	Recommended practice for software specifications development, 4.3.6. Verifiable.
		IEEE Std 1058-1998	Section 4.5.2.2, schedule allocation; Section 4.5.3.2, schedule control plan; and Section 4.5.3.6, metrics collection plan.
		IEEE Std 1045-1992	Sections 6.1 and 8.2.
		IEEE Std 1219-1998	Sections 4 and A.3.5.6.
VE1	The activities for managing the allocated requirements are reviewed with senior management on a periodic basis.	IEEE Std 1028-1997	Section 4, management reviews.
		IEEE Std 1058-1998	Section 4.5.3.5, reporting plan.
		IEEE Std 1219-1998	Sections 4 and A.3.5.6.
		IEEE Std 1074-1997	Annex A, A.1.3, project monitoring and control activities.
VE2	The activities for managing the allocated requirements are reviewed with the project manager on both a periodic and event-driven basis.	IEEE Std 730-2002	Section 4.6.2.8, managerial reviews; and 4.6.3, Other reviews and audits.
		IEEE Std 1058-1998	Section 4.5.3.1, requirements control plan; and Section 4.5.3.5, reporting plan.

IEEE/CMM® Requirements Management Matrix (*continued*)

CMM® Goals and Key Practices	Definiton	IEEE Source	Comments
VE2 (*cont.*)		IEEE Std 1219-1998	Section 4, version description document as output requirement; Section A.5, Forms.
		IEEE Std 1074-1997	Annex A, A.1.3, project monitoring and control activities.
VE3	The software quality assurance group reviews and/or audits the activities and work products for managing the allocated requirements and reports the results.	IEEE Std 730-2002	Section 4.6.
		IEEE Std 1219-1998	Section A.7, software quality assurance.
		IEEE Std 1058-1998	Section 4.5.3.4, quality control plan.

IEEE/CMMI®-SW (Staged) Requirements Management Matrix

CMMI® Goals and Processes		Definition	IEEE Source	Comments
SG1		Requirements are managed and incorporated with project plans and work products are identified.		
	SP 1.1	Obtain an understanding of requirements. Develop an understanding with the requirements providers on the meaning of the requirements.	IEEE Std 830-1998	Section 4.4, joint preparation of the SRS.
			IEEE Std 1233, 1998	Guide for specifications development; Section 4.3.2, communicating to two audiences.
			IEEE Std 1028-1997 (R2002)	Section 4 and Section 5 address management and technical review of requirements, respectively.
			IEEE Std 1074-1997	Section 5, addresses the requirements process; Annex A.2 and A.3.
			ISO/IEC 12207.0-1996	Section 5.3.2.
	SP 1.2	Obtain commitment to requirements. Obtain commitment to the requirements from the project participants.	IEEE Std 1233, 1998	Guide for specifications development.
			IEEE Std 1028-1997 (R2002)	Section 4 and Section 5 address management and technical review of requirements, respectively.
			IEEE Std 1074-1997	Section 5 addresses the requirements process.

IEEE/CMMI®-SW (Staged) Requirements Management Matrix (*continued*)

CMMI® Goals and Processes	Definition	IEEE Source	Comments
SP 1.2 (*cont.*)		ISO/IEC 12207.0-1996	Section 5.3.2., system requirements analysis.
SP 1.3	Manage requirements changes. Manage changes to the requirements as they evolve during the project.	IEEE Std 1028-1997	Provides entry and exit criteria for all procedures for all phases of software life cycle. No requirement in standard for policy statement.
SP 1.4	Maintain bidirectional traceability of requirements. Maintain bi-directional traceability among the requirements and the project plans and work products.	IEEE Std 830-1998	Section 4.3.8, traceable—backward and forward both addressed.
SP 1.5	Identify inconsistencies between project work and requirements. Identify inconsistencies between the project plans and work products and the requirements.	IEEE Std 730-2002	Section A.7, software quality assurance.
		IEEE Std 1219-1998	Section 4.4.
		IEEE Std 1058-1998	Section 4.5.3.4, quality control plan.
GG2	Institutionalize a Managed Process		
GP 2.1 (CO1)	Establish an organizational policy. Establish and maintain an organizational policy for planning and performing the requirements management process.	IEEE Std 1028-1997	Provides entry and exit criteria for all procedures for all phases of software life cycle. No requirement in standard for policy statement.
GP 2.2 (AB1)	Plan the process. Establish and maintain the plan for performing the requirements management process.	IEEE Std 830-1998	Recommended practice for software specifications development, 4.1.
		IEEE Std 1012-1998	Section 7.5, establish specific milestones for initiating and completing each task, for the receipt of each input, and for the delivery of each output.
		ISO/IEC 12207.1-1996	Section 6.22, software requirements specifications.
		IEEE Std 1074-1997	Annex A, A.1.1.3, allocate project resources.

IEEE/CMMI®-SW (Staged) Requirements Management Matrix (*continued*)

CMMI® Goals and Processes	Definition	IEEE Source	Comments
GP 2.2 (*cont.*)		IEEE Std 830-1998	Section 4.8, embedding project requirements in the SRS.
GP 2.3 (AB2)	Provide resources. Provide adequate resources for performing the requirements management process, developing the work products, and providing the services of the process.	IEEE Std 1058-1998	Section 4.5. This section calls out the requirement to allocate resources and funding for activities, but does not reference requirements management specifically.
GP 2.4 (AB3)	Assign responsibility. Assign responsibility and authority for performing the process, developing the work products, and providing the services of the requirements management process.	IEEE Std 1058-1998	Section 4.5.3.1, requirements control plan.
GP 2.5 (AB4)	Train people. Train the people performing or supporting the requirements management process as needed.	IEEE Std 1058-1998	Section 4.5.1.4, project staff training plan; does not specifically address RM training, but should be included in PM plan.
GP 2.6 (DI1)	Manage configurations. Place designated work products of the requirements management process under appropriate levels of configuration management.	IEEE Std 828-1998	Section 4.2, SCM management.
GP 2.7 (DI2)	Identify and involve relevant stakeholders. Identify and involve the relevant stakeholders of the requirements management process as planned.	IEEE Std 1233, 1998	Guide for specifications development.
		IEEE Std 1028-1997 (R2002)	Section 4 and Section 5 address management and technical review of requirements, respectively.
		IEEE Std 1074-1997	Section 5 addresses the requirements process.
		ISO/IEC 12207.0-1996	Section 5.3.2.
GP 2.8 (DI3)	Monitor and control the process. Monitor and control the requirements management process against the plan for performing the process and take appropriate corrective action.	IEEE Std 1028-1997	Section 4, management reviews.
		IEEE Std 1058-1998	Section 4.5.3.5, reporting plan.
		IEEE Std 1219-1998	Sections 4 and A.3.5.6.
		IEEE Std 1074-1997	Annex A, A.1.3, project monitoring and control activities.

IEEE/CMMI®-SW (Staged) Requirements Management Matrix (*continued*)

CMMI® Goals and Processes	Definition	IEEE Source	Comments
GP 2.9 (VE 1)	Objectively evaluate adherence. Objectively evaluate adherence of the requirements management process against its process description, standards, and procedures, and address noncompliance.	IEEE Std 730-2002	Section 4.4.
		IEEE Std 1219-1998	Section A.7, software quality assurance.
		IEEE Std 1058-1998	Section 4.5.3.4, quality control plan.
GP 2.10 (VE2)	Review status with higher-level management. Review the activities, status, and results of the requirements management process with higher-level management and resolve issues.	IEEE Std 1028-1997	Section 4, management reviews.
		IEEE Std 1058-1998	Section 4.5.3.5, reporting plan.
		IEEE Std 1219-1998	Sections 4 and A.3.5.6.
		IEEE Std 1074-1997	Annex A, A.1.3, project monitoring and control activities.

Requirements Management Analysis

In order for the IEEE Std 830-1998 to be an effective instrument for the implementation of Level 2 KPAs, information regarding the management of the defined requirements needs to be added and stated explicitly. Information needs to be included in the SRS, which will demonstrate that the resources, people, tools, funds, and time have been considered for the management of the requirements that are defined.

This information should relate directly to the Software Project Management Plan (SPMP). For example, if one person is required to perform the duties of the requirements manager, this should be identified in the SPMP. This section could reference the SPMP if the information is contained in that document. However, the present IEEE Standard for SPMPs does not require the identification of requirements management issues such as resources and funding, reporting procedures, and training.

Example of IEEE KPA Support for Requirements Management

IEEE Standards 830 and 1233 both provide dramatic, direct support for CMM®/CMMI® Level 2 requirements. IEEE Std 830, Recommended Practice for Software Requirements Specifications, details what is needed to create a sound software requirements specification, including the provision of documentation templates. IEEE Std 1233 provides specific guidance on how to elicit, define, and communicate systems requirements effectively. The following provides an example of the type of direct KPA support found within IEEE Std 830.

> CMM® AC1. The software engineering group reviews the allocated requirements before they are incorporated into the software project.

CMMI®-SW (Staged) SP1.1. Obtain an understanding of requirements; Develop an understanding with the requirements providers on the meaning of the requirements.

IEEE Std 830—4.4 Joint Preparation of the SRS. The software development process should begin with supplier and customer agreement on what the completed software must do. This agreement, in the form of an SRS, should be jointly prepared. This is important because usually neither the customer nor the supplier is qualified to write a good SRS alone:

a) Customers usually do not understand the software design and development process well enough to write a usable SRS.
b) Suppliers usually do not understand the customer's problem and field of endeavor well enough to specify requirements for a satisfactory system.

Therefore, the customer and the supplier should work together to produce a well-written and completely understood SRS.

Requirements Traceability

Requirements traceability is a CMM® cornerstone for achieving Level 2 and the traceability throughout the defined lifecycle can be accomplished by the addition of a traceability matrix. The example in Table 5-1 supports backward and forward traceability for validation testing. Additional columns should be added to support traceability to software, or system, design and development. The conversion of this type of matrix to a database tracking system is a common practice for developing organizations. There are many commercially available tools that support requirements tracking.

Change Enhancement Requests

The identification and tracking of application requirements is accomplished through some type of change enhancement request (CER). A CER is used to control changes to all documents and software under SCM control and for documents and software that have been released to SCM. Ideally, the CER is an on-line form that the software developers use to submit changes to software and documents to the SCM Lead for updating the product. Table 5-2 provides an example of the types of data that may be captured and tracked in support of application development or modification.

Table 5-1 Requirements traceability matrix example

CER #	Requirement Name	Priority	Risk	Requirements Document Paragraph	Validation Method(s)	Formal Test Paragraph	Status
1	Assignments and Terminations Module (A&T) Performance	2	M	3.1.1.1.1.1	Test Inspection Demonstration	4.1.2.3 4.1.2.6 4.1.2.7 4.1.2.9	Open

Table 5-2 Typical elements in a CER

Element	Values
CER #	Unique CER identifier assigned by CRC.

Type	May be one of the following:
	    BCR—Baseline change request. CER indicating additional/changed requirement.
	    SPR—Software problem report. CER indicating software problem identified externally during or after beta test.
	    ITR—Internal test report. CER indicating software problem identified internally through validation procedures.
	    DOC—Documentation change. CER indicting change to software documentation.

Status	One of the following:
	    OPEN—CER in queue for work assignment.
	    TESTING—CER in validation test.
	    WORKING—CER assignment for implementation as SCR(s).
	    VOIDED—CER deemed not appropriate for software release.
	    HOLD—CER status to be determined.
	    FIXED—CER incorporation into software complete.

Category	One of the following:
	    DATA—Problem resulting from inaccurate data processing.
	    DOC—Documentation inaccurate.
	    REQT—Problem caused by inaccurate requirement

Priority	One of the following:
	    1. Highest priority; indicates software crash with no work-around.
	    2. Cannot perform required functionality; available work around.
	    3. Lowest priority; not critical to software performance.
	    0. CER has no established priority.

Date Submitted	The date CER initiated; defaults with current date.
Date Closed	The date CER passes Module test.
Originator	Identifies the source of CER.
Problem Description	Complete and detailed description of the problem, enhancement, or requirement.
Short Title	Short title summarizing CER.
SRS Ref #	Cross reference to associated software requirements specification (SRS) paragraph identifier.
SDD Ref #	Cross reference to associated software design document (SDD) paragraph identifier(s).
STP Ref #	Cross reference to associated software test plan (STP) paragraph identifier(s).
Targeted Version	Software module descriptor and version number; determined by SCM lead.
File	Files affected by CER.
Functions Affected	List of functions affected by CER.
Time Estimated	Estimated hours until implementation complete.
Time Actual	Actual hours to complete CER implementation.

(continued)

Table 5-2 Typical elements in a CER

Element	Values
Module	Name of module affected by CER.
Problem Description	Description of original CER; add comments regarding implementation.
Release	Release(s) affected by CER.
CER #	Cross-reference number used to associate item with other relevant CERs.
Software Engineering Name	Last name of engineer implementing CER.

The modification of the recommended SRS table of contents to support the goals of the CMM®/CMMI®-SW (Staged) methodologies more directly could look like the one shown in Figure 5-1.

The modification of the recommended SysRS table of contents to support the goals of the CMM®/CMMI®-SW (Staged) methodologies more directly could look like the one shown in Figure 5-2.

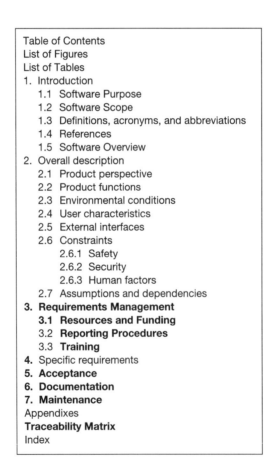

Table of Contents
List of Figures
List of Tables
1. Introduction
   1.1 Software Purpose
   1.2 Software Scope
   1.3 Definitions, acronyms, and abbreviations
   1.4 References
   1.5 Software Overview
2. Overall description
   2.1 Product perspective
   2.2 Product functions
   2.3 Environmental conditions
   2.4 User characteristics
   2.5 External interfaces
   2.6 Constraints
      2.6.1 Safety
      2.6.2 Security
      2.6.3 Human factors
   2.7 Assumptions and dependencies
3. **Requirements Management**
   3.1 Resources and Funding
   3.2 Reporting Procedures
   3.3 Training
4. Specific requirements
5. **Acceptance**
6. **Documentation**
7. **Maintenance**
Appendixes
Traceability Matrix
Index

Figure 5-1 Modified SRS format (IEEE Std 830-1998).

```
Table of Contents
List of Figures
List of Tables
1. Introduction
    1.1  System Purpose
    1.2  System Scope
    1.3  Definitions, acronyms, and abbreviations
    1.4  References
    1.5  System Overview
2. General System Description
    2.1  System context
    2.2  System modes and states
    2.3  Major system capabilities
    2.4  Major system conditions
    2.5  Major system constraints
    2.6  User characteristics
    2.7  Assumptions and dependencies
    2.8  Operational scenarios
3. System capabilities, conditions, and constraints
    3.1  Physical
        3.1.1  Construction
        3.1.2  Durability
        3.1.3  Adaptability
        3.1.4  Environmental conditions
    3.2  System performance characteristics
    3.3  System security
    3.4  Information management
    3.5  System operations
        3.5.1  System human factors
        3.5.2  System maintainability
        3.5.3  System reliability
    3.6  Policy and regulation
    3.7  System life cycle sustainment
4. System Interfaces
8. **Requirements Management**
    8.1  **Resources and Funding**
    8.2  **Reporting Procedures**
    8.3  **Training**
9. Specific requirements (several different are provided for organizing this section.)
Appendixes
**Traceability Matrix**
Index
```

Figure 5-2   Modified SysRS format (IEEE Std 1233-1998).

The Goals for CMM® Requirements Management Revisited

SW-CMM® Goals for Software Requirements Management. Controlling requirements is the most crucial step in ensuring that Level 1 processes are repeatable. The CMM®/CMMI® KPAs focus on the management of requirements rather than on requirements engineering. IEEE Standards support both concepts.

Goal 1. System requirements allocated to software are controlled to establish a baseline for software engineering and management use. Goal 1 contains the phrase "system requirements allocated to software." This phrase is used intentionally to cover the most general case of software activity, one in which software is only a part of a large system with many hardware components that do not execute software. For smaller systems, those with no additional hardware requirements, the system requirements may be considered to be the same as the software requirements.

Goal 1 has to do with establishing the requirements baseline. The requirements must be reviewed and approved by the software engineering group and placed in a baseline. IEEE Standards 830, 1233, and 1219 directly support this goal and are invaluable resources for the documentation of initial software requirements baselines.

Goal 2. Software plans, products, and activities are kept consistent with the system requirements allocated to software. Goal 2 anticipates the changes to requirements and requires other aspects of the project to conform, ensuring that all project plans, products, and activities are kept consistent with any changes in software requirements. This goal focuses on how software requirements are managed during the project lifecycle. IEEE Standards 1058, 1028, 828, and 730 will help any organization define the "how to" of their requirements management.

CMMI®-SW (Staged) Goals for Requirements Management. The CMMI®-SW (Staged) emphasizes bidirectional traceability, elevating it to specific practice. The support provided by IEEE software engineering standards for requirements traceability was previously addressed when describing the IEEE software engineering support provided to the SW-CMM® software-requirements management KPA.

The CMMI® requires that a plan for performing the requirements management process be established and that all stakeholders be identified according to this plan. IEEE Standard 830 provides detailed guidance in support of the development of a software requirements management plan. This plan provides detail on what is required to effectively manage software requirements and how to document these requirements in a management plan.

SG1 Manage Requirements. Requirements are managed and incorporated with project plans and work products are identified. As previously stated, in support of the CMM® Goal 1, specific support for this goal is provided by IEEE Standards 830, 1233, and 1219. The IEEE software engineering standards set support the idea of requirements-driven software development and/or maintenance activities.

GG2 Institutionalize a Managed Process. The process is institutionalized as a managed process. The implementation of IEEE Std 830, Recommended Practice for Software Requirements Specifications, assures the development of an SRS but does not support the requirement of the institutionalization of software requirements management. At a minimum, IEEE Std 830 must be used in conjunction with IEEE Std 1058, Standard for Software Project Management Plans, in support of this goal. Incorporating requirements management into the project planning and management helps to ensure that it is integrated into the lifecycle of the project.

SOFTWARE PROJECT PLANNING

The purpose of CMM® Level 2 Software Project Planning is to establish reasonable plans for performing software engineering and software project management. The IEEE Standard for Project Management Plans (IEEE Std 1058) can be used as a model for this CMM® Level 2 process. IEEE Std 1058 specifies a suggested format for a project management plan. This document may be used as a guide for documenting the practices and procedures unique to each organization for all types of software efforts.

A software project management plan has three main components: the work to be done, the resources with which to do it, and the money to pay for it. IEEE Std 1058 addresses each of these components in detail. The software development plan provides the basis for performing and managing the software project's activities.

The Goals for CMM® Software Project Planning

This KPA contains 25 practices, more than any other KPA. The goals associated with this KPA address the size of the software effort, the plan to carry out the effort, necessary commitments, and the agreement to these commitments:

Goal 1. Software estimates are documented for use in planning and tracking the software project.
Goal 2. Software project activities and commitments are planned and documented.
Goal 3. Affected groups and individuals agree to their commitments related to the software project.

Project management involves developing estimates for all work to be performed; establishing the necessary commitments in support of the work; defining the plan to perform the work; tracking and reviewing the software accomplishments and results against these documented estimates, commitments, and plans, and adjusting plans based on the actual accomplishments and results.

Software planning begins with a statement of the work to be performed and the supporting requirements. The software planning process includes size estimation of the software work products and the resources needed, production of a schedule, identification and assessment of software risk, and negotiation of commitments. Iteration through these steps may be necessary in order to establish a plan for the software project.

The software plan (development or maintenance) provides the basis for performing and managing the software project's activities and addresses the commitments to the software project's customer according to the resources, constraints, and capabilities of the software project. This plan is also used as the basis for tracking the software activities and communicating status.

Progress is primarily determined by comparing the actual software size, effort, cost, and schedule to the plan at selected milestones. When it is determined that the software project's plans are not being met, corrective actions are taken. These actions may include revision of the software plan to reflect the actual accomplishments, replanning the remaining work, or taking actions to improve performance.

The Goals for CMMI®-SW (Staged) Project Planning

The CMMI®-SW (Staged) requires that the project planning activities must reflect all items as described in the SW-CMM® Software Project Planning KPA. However, the CMMI® requires that changes to all available and estimated resources must be reflected in any project planning activities. This is implied in the SW-CMM®, but is specifically addressed in the CMMI-SW®:

SG1. Establish Estimates. Estimates of project planning parameters are established and maintained.

SG2. Develop a project Plan. A project plan is established and maintained as the basis for managing the project.

SG3. Obtain commitment to the plan. Commitments to the project plan are established and maintained.

GG2. Institutionalize a managed process. The process is institutionalized as a managed process.

The CMMI®-SW has a requirement to establish and maintain a plan that describes the performance of the project planning process. IEEE Standard 1058 provides support for this KPA and can be used to support the development of a plan that provides the detailed description of this process.

Supporting IEEE Software Engineering Standards

IEEE Software Project Management Plan IEEE Std 1058. There are distinct advantages to following the IEEE Software Project Management Plan (SPMP) standard. First, it is a standard drawn up by representatives of many organizations involved in software development. Input is received from industry and universities, and the members of the working group and reviewing teams. This collection of individuals represents many years of SPMP development experience. Second, the IEEE SPMP is designed for use with all types of software products, irrespective of size. It does not impose a specific process model or prescribe specific techniques. The plan basically provides a framework for organizations to follow. By adhering to this framework, organizations reap the benefits of standardization.

The project plan should reflect the separate phases of development with its major components being the deliverables, the milestones, and budget. The plan should include aspects such as the life-cycle model to be used, the organizational structure of the development organization, responsibilities, objectives and priorities, techniques and CASoftware Engineering tools used, detailed schedules, budgets, and resource allocations.

The following matrix provides a cross-reference of the SEI CMM® Level 2 Software Project Planning KPA to relevant supporting documentation from the IEEE Software Engineering Standards Collection. This is not meant to be an exhaustive list, rather is used to illustrate how the various KPAs are directly supported by IEEE Standards.

IEEE/CMM® Software Project Planning Matrix

CMM® Goals and Key Practices	Definiton	IEEE Source	Comments
G1	Software estimates are documented for use in planning and tracking the software project.		
AC9	Estimates for the size of the software work products (or changes to the size of software work products) are derived according to a documented procedure.	IEEE Std 830-1998	Section 4.3.8, requirements should be traceable—supported if size estimates are part of the initial requirement.
		IEEE Std 1058-1998	Section 4.5, SPMP document estimates and how they were derived.
		IEEE Std 1219-1998	Section 4, maintenance size estimates.
		IEEE Std 12207.1-1996	Section 6, specific information item content guidelines.
AC10	Estimates for the software project's effort and costs are derived according to a documented procedure.	IEEE Std 1058-1998	Section 4.5, SPMP document estimates and how they were derived.
		IEEE Std 1045-1992	Can use the examples provided in this plan
		IEEE Std 1219-1998	Section 4, requirement for estimate, but no detail on how to derive these.
AC11	Estimates for the project's critical computer resources are derived according to a documented procedure.	IEEE Std 1058-1998	Section 4.5, SPMP documents require the documentation of critical computer resources, and also the software tools, special testing and simulation facilities, and administration support required.
		IEEE Std 1219-1998	Section A.3.5.5, resource allocation
AC12	The project's software schedule is derived according to a documented procedure.	IEEE Std 1058-1998	Section 4.1.1.4, schedule and budget summary.
		IEEE Std 1219-1998	Section A3.
AC15	Software planning data are recorded.	IEEE Std 1058-1998	Section 4.
		IEEE Std 1219-1998	Section A3.
		IEEE Std 12207.1-1996	Section 6, specific information item content guidelines.
G2	Software project activities and commitments are planned and documented.		

IEEE/CMM® Software Project Planning Matrix (*continued*)

CMM® Goals and Key Practices	Definiton	IEEE Source	Comments
AC2	Software project planning is initiated in the early stages of, and in parallel with, the overall project planning.	IEEE Std 1220-1998	Section 5.6, simultaneous engineering of life cycle processes.
AC5	A software life cycle with predefined stages of manageable size is identified or defined.	IEEE Std 1058-1998	Section 6.1, process model.
		IEEE Std 1074-1997	How to develop software life cycle processes.
		IEEE Std 1220-1998	Section 5.6.1, requirement for life cycle process development.
AC6	The project's software development plan is developed according to a documented procedure.	IEEE Std 1058-1998	Derived from this IEEE standard for software project management plans.
		IEEE Std 1219-1998	Section A.3.5, develop maintenance plan.
AC7	The plan for the software project is documented.	IEEE Std 1058-1998	Software management plan format.
		IEEE Std 1219-1998	Software maintenance plan format.
		IEEE Std 12207.1-1996	Section 6, specific information item content guidelines.
AC8	Software work products that are needed to establish and maintain control of the software project are identified.	IEEE Std 1058-1998	Section 4.5.1.3, resource acquisition; Section 4.5.2.1, work activities; Section 4.6.2, methods, tools, and techniques.
		IEEE Std 1219-1998	Section A.3.5.5, resource allocation.
AC13	The software risks associated with the cost, resource, schedule, and technical aspects of the project are identified, assessed, and documented.	IEEE Std 1058-1998	Section 4.5.3.6 of this document or Section 5 of SPMP.
		IEEE Std 1219-1998	Software maintenance. Sections 4 and A.8, risk assessment.
		IEEE Std 730-2002	Section 4.15 (Section 15 of SQAP), risk management.
AC14	Plans for the project's software engineering facilities and support tools are prepared.	IEEE Std 1058-1998	Section 4.6.3, infrastructure plan.
		IEEE Std 829-1998	Forms for recording, tracking, and implementing software maintenance.
		IEEE Std 12207.1-1996	Section 6, specific information item content guidelines.

IEEE/CMM® Software Project Planning Matrix (*continued*)

CMM® Goals and Key Practices	Definiton	IEEE Source	Comments
G3	Affected groups and individuals agree to their commitments related to the software project.		
AC1	The software engineering group participates on the project proposal team.	IEEE Std 1058-1998	Section 4.4.3 states that all roles and responsibilities should be defined but the requirement for the software engineering group to participate on the project proposal team is not specifically called out.
		IEEE Std 1219-1998	Section 4, supports initial analyst review of modification request.
AC3	The software engineering group participates with other affected groups in the overall project planning throughout the project's life.	IEEE Std 1058-1998	Section 4.5.3.5, reporting plan requires identification of affected groups and reporting methods.
		IEEE Std 1042-1987	Identifies requirements for configuration control board.
		IEEE Std 830-1998	Section 4.4, joint preparation of the SRS.
		IEEE Std 1220-1998	Section 4.12, calls out types of reviews but does not designate participants.
AC4	Software project commitments made to individuals and groups external to the organization are reviewed with senior management according to a documented procedure.	IEEE Std 1058-1998	Section 4.1.1.3, project deliverables.
CO1	A project software manager is designated to be responsible for negotiating commitments and developing the project's software development plan.	IEEE Std 1058-1998	Section 4.5.1.2, staffing plan.
CO2	The project follows a written organizational policy for planning a software project.	IEEE Std 1058-1998	Could be derived from this IEEE standard for software project management plans.
		IEEE Std 1219-1998	Could be derived from this IEEE standard for software maintenance.
			No requirement in standard for policy statement.
AB1	A documented and approved statement of work exists for the software project.	IEEE Std 1058-1998	Section 4.1.1.3, contract data requirements list (CDRL); or Section 4.5.2, work plan.

IEEE/CMM® Software Project Planning Matrix (*continued*)

CMM® Goals and Key Practices	Definiton	IEEE Source	Comments
AB2	Responsibilities for developing the software development plan are assigned.	IEEE Std 1058-1998	Section 4.5.2.2 and 4.5.2.3.
AB3	Adequate resources and funding are provided for planning the software project.	IEEE Std 1058-1998	Section 4.1.1.2, assumptions and constraints; and Section 4.1.1.4, schedule and budget summary.
AB4	The software managers, software engineers, and other individuals involved in the software project planning are trained in the software estimating and planning procedures applicable to their areas of responsibility.	IEEE Std 1058-1998	Section 4.5.1, project start-up plan requires the identification of required training; it is up to the PM to include this type of training in the plan.
		IEEE Std 1045-1992 And IEEE 1061-1998	A training document could be developed from these IEEE standards to meet this CMM® ability.
ME1	Measurements are made and used to determine the status of the software planning activities.	IEEE Std 1061-1998	This standard could be used to develop a metrics plan that would identify metrics captured regarding software planning activities, but these types of metrics are not specifically identified.
		IEEE Std 1058-1998	Section 4.5.3.6 states that the methods, tools, and techniques used for project metrics collection and retention be specified, but software planning metrics are not specifically identified.
		IEEE Std 1219-1998	Section A.3.5.6 states that once a process is in place it should be measured and tracked, but offers no specifics.
VE1	The activities for software project planning are reviewed with senior management on a periodic basis.	IEEE Std 1058-1998	Section 4.5.3.3, budget control plan; Section 4.5.3.5, reporting plan; and Section 4.5.4, risk management plan.
		IEEE Std 1219-1998	Section 4.1.3, control. Specifies the process of analyzing the maintenance request. Does not specifically identify level of personnel to be involved.
VE2	The activities for software project planning are reviewed with the project manager on	IEEE Std 1058-1998	Section 4.5.3.5, reporting plan.

IEEE/CMM® Software Project Planning Matrix (continued)

CMM® Goals and Key Practices	Definiton	IEEE Source	Comments
	both a periodic and event-driven basis.	IEEE Std 730-2002	Section 4.6, activities for joint review and reporting mechanisms and Section 4.3.2, tasks.
VE3	The software quality assurance group reviews and/or audits the activities and work products for software project planning and reports the results.	IEEE Std 1058-1998	Section 4.5.3.4, quality control plan.
		IEEE Std 730-2002	Section 4.6, activities for audits, problem resolution, and reporting.

IEEE/CMMI®-SW (Staged) Project Planning Matrix

CMMI® Goals and Processes	Definition	IEEE Source	Comments
SG1	Establish estimates:		
SP 1.1	Estimate the scope of the project. Establish a top-level work breakdown structure (WBS) to estimate the scope of the project.	IEEE Std 1058-1998	Section 4.1.1.1, purpose, scope, and objectives; Section 4.1.1.4, schedule and budget summary.
SP 1.2	Establish estimates of work product and task attributes. Establish and maintain estimates of the attributes of the work products and tasks.	IEEE Std 1058-1998	Section 4.5.1.1, estimation plan; Section 4.1.1.3, project deliverables.
SP 1.3	Define project life cycle. Define the project life-cycle phases upon which to scope the planning effort.	IEEE Std 1058-1998	Section 6.1, process model.
		IEEE Std 1074-1997	How to develop software life-cycle processes.
		IEEE Std 1220-1998	Section 5.6.1, requirement for life-
SP 1.4	Determine estimates of effort and cost. Estimate the project effort and cost for the work products and tasks based on estimation rationale.	IEEE Std 1058-1998	Section 4.5, SPMP document estimates and how they were derived.
		IEEE Std 1045-1992	Can use the examples provided in this plan.
		IEEE Std 1219-1998	Section 4, requirement for estimate, but no detail on how to derive these.
SG2	Develop a project plan.		
SP 2.1	Establish the budget and schedule. Establish and maintain the project's budget and schedule.	IEEE Std 1058-1998	Section 4.1.1.4, schedule and budget summary.
		IEEE Std 1219-1998	Section A3.

IEEE/CMMI®-SW (Staged) Project Planning Matrix (*continued*)

CMMI® Goals and Processes	Definition	IEEE Source	Comments
		IEEE Std 12207.1-1996	Section 6, specific information item content guidelines.
SP 2.2	Identify project risks. Identify and analyze project risks.	IEEE Std 1058-1998	Section 4.5.3.6 of this document or Section 5 of SPMP.
		IEEE Std 1219-1993	Software maintenance. Section 3 and A.8, risk assessment.
		IEEE Std 730-2002	Section 4.15 (Section 15 of SQAP), risk management.
SP 2.3	Plan for data management. Plan for the management of project data.	IEEE Std 1058-1998	Section 4.
		IEEE Std 1219-1998	Section A3.
		IEEE Std 12207.1-1996	Section 6, specific information item content guidelines.
SP 2.4	Plan for project resources. Plan for necessary resources to perform the project.	IEEE Std 1058-1998	Section 4.5 SPMP documents require the documentation of critical computer resources, and also the software tools, special testing and simulation facilities, and administrative support required.
		IEEE Std 1219-1998	Section A.3.5.5, resource allocation.
		IEEE Std 12207.1-1996	Section 6, specific information item content guidelines.
SP 2.5	Plan for needed knowledge and skills. Plan for knowledge and skills needed to perform the project.	IEEE Std 1058-1998	Section 4.5.1, project start-up plan requires the identification of required training; it is up to the PM to include this type of training in the plan.
		IEEE Std 1045-1992 And IEEE 1061-1998	A training document could be developed from these IEEE standards to meet this CMM® ability.
SP 2.6	Plan stakeholder involvement. Plan the involvement of identified stakeholders.	IEEE Std 1058-1998	Section 4.1.1.3, project deliverables; Section 4.4.3 states that all roles and responsibilities should be defined but the requirement for the Software Engineering group to participate on the project proposal team is not specifically called out; Section 4.5.3.5, re-

IEEE/CMMI®-SW (Staged) Project Planning Matrix (*continued*)

CMMI® Goals and Processes	Definition	IEEE Source	Comments
SP 2.6 (*cont.*)			porting plan requires identification of affected groups and reporting methods.
		IEEE Std 1219-1998	Section 4, supports initial analyst review of modification request.
		IEEE Std 1042-1987	Identifies requirements for configuration control board.
		IEEE Std 830-1998	Section 4.4, joint preparation of the SRS.
		IEEE Std 1220-1998	Section 4.12 Calls out types of reviews but does not designate participants.
SP 2.7	Establish the project plan. Establish and maintain the overall project plan content.	IEEE Std 1058-1998	Section 4.1.2, evolution of the software project management plan.
		IEEE Std 12207.1-1996	Section 6, specific information item content guidelines.
SG3	Obtain commitment to the plan.		
SP 3.1	Review plans that affect the project. Review all plans that affect the project to understand project commitments.	IEEE Std 1058-1998	Section 4.5.3.3, budget control plan; Section 4.5.3.5, reporting plan; Section 4.5.4, risk management plan; Section 4.5.3.4, quality control plan.
		IEEE Std 1219-1998	Section 4, control. Specifies the process of analyzing the maintenance request. Does not specifically identify level of personnel to be involved.
		IEEE Std 730-2002	Section 4.6, activities for joint review and reporting mechanisms; and Section 4.3.2, tasks.
SP 3.2	Reconcile work and resource levels. Reconcile the project plan to reflect available and estimated resources.	IEEE Std 1058-1998	Section 4.5.1.3, resource acquisition; Section 4.5.2.1, work activities; Section 4.6.2, methods, tools, and techniques.
		IEEE Std 1219-1998	Section A.3.5.5, resource allocation.
SP 3.3	Obtain plan commitment. Obtain commitment from relevant stakeholders responsible for performing and supporting plan execution.	IEEE Std 1058-1998	Section 4.5.3.3, budget control plan; Section 4.5.3.5, reporting plan; Section 4.5.4, risk management plan; Section 4.5.3.4, quality control plan.
		IEEE Std 1219-1998	Section 4, control. Specifies the process of analyzing the maintenance request. Does not specifically identify level of personnel to be involved.

IEEE/CMMI®-SW (Staged) Project Planning Matrix (continued)

CMMI® Goals and Processes	Definition	IEEE Source	Comments
		IEEE Std 730-2002	Section 4.6, activities for joint review and reporting mechanisms; and Section 4.3.2, tasks.
GG2	Institutionalize a managed process.		
GP 2.1 (CO1)	Establish an organizational policy. Establish and maintain an organizational policy for planning and performing the project planning process.	IEEE Std 1058-1998	Could be derived from this IEEE standard for software project management plans.
		IEEE Std 1219-1998	Could be derived from this IEEE standard for software maintenance.
			No requirement in standard for policy statement.
GP 2.2 (AB1)	Plan the process. Establish and maintain the plan for performing the project planning process.	IEEE Std 1058-1998	Software management plan format.
		IEEE Std 1219-1998	Software maintenance plan format.
		IEEE Std 12207.1-1996	Section 6, specific information item content guidelines.
GP 2.3 (AB 2)	Provide resources. Provide adequate resources for performing the project planning process, developing the work products, and providing the services of the process.	IEEE Std 1058-1998	Section 4.1.1.2, assumptions and constraints; and Section 4.1.1.4, schedule and budget summary.
GP 2.4 (AB3)	Assign responsibility. Assign responsibility and authority for performing the process, developing the work products, and providing the services of the project planning process.	IEEE Std 1058-1998	Sections 4.5.2.2 and 4.5.2.3
GP 2.5 (AB4)	Train people. Train the people performing or supporting the project planning process as needed.	IEEE Std 1058-1998	Section 4.5.1.4, project staff training plan; does not specifically address training in support of project planning, but should be included in PM plan.
GP 2.6 (DI1)	Manage configurations. Place designated work products of the project planning process under appropriate levels of configuration management.	IEEE Std 1058-1998	Section 4.7.1, configuration management plan.
		IEEE Std 828-1998	Standard for software configuration management plans.
GP 2.7 (DI2)	Identify and involve relevant stakeholders. Identify and involve the relevant stakeholders of the project planning process as planned.	IEEE Std 1058-1998	Section 4.1.1.3, project deliverables; Section 4.4.3 states that all roles and responsibilities should be defined but the requirement for the software engineering group to participate on the project proposal team is not specifical-

IEEE/CMMI®-SW (Staged) Project Planning Matrix (*continued*)

CMMI® Goals and Processes	Definition	IEEE Source	Comments
GP 2.7 (DI2) (*cont.*)			ly called out; Section 4.5.3.5, reporting plan, requires identification of affected groups and reporting methods.
		IEEE Std 1219-1998	Section 4, supports initial analyst review of modification request.
		IEEE Std 1042-1987	Identifies requirements for configuration control board.
		IEEE Std 830-1998	Section 4.4, joint preparation of the SRS.
		IEEE Std 1220-1998	Section 4.12, calls out types of reviews but does not designate participants.
GP 2.8 (DI3)	Monitor and control the process. Monitor and control the project planning process against the plan for performing the process and take appropriate corrective action.	IEEE Std 1061-1998	This standard could be used to develop a metrics plan that would identify metrics captured regarding software planning activities, but these types of metrics are not specifically identified.
		IEEE Std 1058-1998	Section 4.5.3.6 states that the methods, tools, and techniques used for project metrics collection and retention be specified, but software planning metrics are not specifically identified.
		IEEE Std 1219-1998	Section A.3.5.6 states that once process is in place, should be measured and tracked, but offers no specifics.
GP 2.9 (VE 1)	Objectively evaluate adherence. Objectively evaluate adherence of the project planning process against its process description, standards, and procedures, and address noncompliance.	IEEE Std 1058-1998	Section 4.5.3.4, quality control plan.
		IEEE Std 730-2002	Section 4.6, activities for audits, problem resolution, and reporting.
GP 2.10 (VE2)	Review status with higher-level management. Review the activities, status, and results of the project planning process with higher-level management and resolve issues.	IEEE Std 1058-1998	Section 4.5.3.5, reporting plan requires identification of affected groups and reporting methods.
		IEEE Std 1042-1987	Identifies requirements for configuration control board.
		IEEE Std 830-1998	Section 4.4, Joint preparation of the SRS.
		IEEE Std 1220-1998	Section 4.12, calls out types of reviews but does not designate participants.

IEEE Standard for Software Test Documentation IEEE Std 829. The IEEE Standard for Software Test Documentation, IEEE 829, describes the requirements for test documentation. This includes the specifications for a test plan, associated test cases, and test summary reporting. For each individual document, the standard describes the structure and content. Additionally, an example of each type of test document, including implementation and usage guidelines are provided in the annex. This standard does not recommend specific test design techniques or tools. This standard covers a broad range of test documents, from the test plan and test design specification to the test summary report.

IEEE Standard for Software Maintenance IEEE Std 1219. IEEE Std 1219 provides detailed guidance in support of the requirements the planning, management, execution, and documentation of software maintenance activities. The basic process model described supports the input, process, output, and control process model for software maintenance. Information is provided regarding the types of metrics used in support of software maintenance activities with no predisposition of lifecycle model (e.g., incremental, spiral). This standard also provides specific guidelines for maintenance plan development.

Project Planning Analysis

IEEE std 1058-1998 is an effective instrument for the implementation of Level 2 KPAs. However, information regarding the metrics required in support of software project planning activities needs to be added and stated explicitly. It is also important to note that whereas the various Level 2 KPAs are addressed by the major headings of IEEE 1058, the details required to support the CMM® can be lost if each section is not carefully addressed while bearing the specific CMM® project planning commitments, abilities, measurements, verification and activities in mind.

The modification of the recommended SPMP table of contents to support the goals of the CMM® more directly is shown in Figure 5-3.

Example of IEEE KPA Support for Software Project Planning

> CMM® AC10. Estimates for the software project's effort and costs are derived according to a documented procedure.
> CMMI®-SW (Staged) SP1.4. Determine estimates of effort and cost. Estimate the project effort and cost for the work products and tasks based on estimation rationale.
> IEEE Std 1058 Section 4.5 Managerial Process Plans. 4.5.1.1 Estimation plan (Subclause 5.1.1 of the SPMP). This subclause of the SPMP shall specify the cost and schedule for conducting the project as well as methods, tools, and techniques used to estimate project cost, schedule, resource requirements, and associated confidence levels. In addition, the basis of estimation shall be specified to include techniques such as analogy, rule of thumb, or local history and the sources of data. This subclause shall also specify the methods, tools, and techniques that will be used to periodically reestimate the cost, schedule, and resources needed to complete the project. Reestimation may be done on a monthly basis and periodically as necessary.

Table of Contents
1. Overview
    1.1 Project Summary
        1.1.1 Purpose, scope, and objectives
        1.1.2 Assumptions and constraints
        1.1.3 Project deliverables
        1.1.4 Schedule and budget summary
    1.2 Evolution of the plan
2. References
3. Definitions
4. Project organization
    9.1 **Organizational Policies**
    9.2 External interfaces
    9.3 Internal structure
    9.4 Roles and responsibilities
10. Managerial process
    10.1 Start-up
        10.1.1 Estimation
        10.1.2 Staffing
        10.1.3 Resource acquisition
    10.2 Work plan
        10.2.1 Work activities
        10.2.2 Schedule allocation
        10.2.3 Resource allocation
        10.2.4 Budget allocation
        10.2.5 Reporting
        10.2.6 Metrics
            10.2.6.1 Planning
            10.2.6.2 Execution
    10.3 Closeout plan
11. Technical process
    11.1 Process model
    11.2 Methods, tools, and techniques
    11.3 Infrastructure
12. Supporting process plans
    12.1 Configuration management plan
    12.2 Verification and validation plan
    12.3 Documentation plan
    12.4 Quality assurance plan
        12.4.1 Reviews and audits
        12.4.2 Problem resolution plan
    12.5 Risk management plan
    12.6 Software Requirements Specification
    12.7 Subcontractor management plan
    12.8 Process improvement plan
    12.9 Metrics plan
    12.10 Staff training plan
13. Additional plans
Annexes
Index

Figure 5-3 Example SPMP.

The Goals of Software Project Planning Revisited

SW-CMM® Goals for Software Project Planning. These goals support the effective management of a software project. They define the technical and managerial processes required in support of the development of software products that satisfy customer requirements. IEEE Standards supplement these goals by providing documented guidance on effective software project management.

> Goal 1. Software estimates are documented for use in planning and tracking the software project.
> Goal 2. Software project activities and commitments are planned and documented.
> Goal 3. Affected groups and individuals agree to their commitments related to the software project.

These goals require that the project size, schedule, and effort be examined and documented. They also require that estimates be documented and used in support of project tracking. Also required is the agreement of those involved in the project to their commitments. Without this agreement, the plan, estimates, and schedule are not likely to function. IEEE Standards 1058 primarily support the bulk of the requirements associated with software project planning. IEEE Std 1045 provides examples of software productivity metrics. IEEE Std 730 supports risk definition and tracking, and audits and reviews.

CMMI®-SW (Staged) Goals for Project Planning. The SW-CMM® identifies risk management as a key practice in software project planning, software project tracking and oversight, and integrated software management. At Level 2, projects identify, assess, document, and track software risks. In practice, many current software projects fail to identify project-specific risk and mitigation and do not effectively manage risk throughout the project life cycle.

> SG 1. Establish estimates. Estimates of project planning parameters are established and maintained.
> SG 2. Develop a project plan. A project plan is established and maintained as the basis for managing the project.
> SG 3. Obtain commitment to the plan. Commitments to the project plan are established and maintained.
> GG 2. Institutionalize a managed process. The process is institutionalized as a managed process.

The CMMI®-SW elevates risk management to a process area, with its own set of required goals and practices as they apply to software. In specifies that as a specific project activity, risk management must be institutionalized through written organizational policies, plans, allocated resources, assigned responsibilities, training, configuration management of work products, quality audits, and management reviews.

SOFTWARE PROJECT TRACKING AND OVERSIGHT

Simply initially estimating the duration and total cost of a software effort is not sufficient. Planning must continue throughout the software development and maintenance process.

Project monitoring (tracking) and control of the management process encompasses most of the development process, providing visibility into project activities and status.

This includes all activities that project management has to perform to ensure that the project objectives are met and that development proceeds according to the plan. As cost, schedule, and quality are the key issues for any project, most activity revolves around monitoring factors that affect these.

Monitoring potential risks for the project, which might prevent completion, is another important management activity. If the information obtained by this oversight suggests that significant risks exist, corrective action may be taken.

For effective project monitoring, the information evaluated should be objective and quantitative. Each of the five Level 2 CMM® KPAs has one or more measurement and analysis requirements. This measurement information is emphasized in the Software Project Tracking and Oversight KPA. The purpose of software project tracking and oversight is to provide insight into actual progress of a software effort so that management can take effective actions when the software project's performance deviates from the documented software project management plan.

The Goals of Software Project Tracking and Oversight

The group of practices under this KPA provides visibility into activities and status of a software project. The point of this is that if you have this visibility, then when a deviation occurs corrective action can be taken. This KPA really focuses on the items that indicate whether the processes in place can repeatedly be used for effective project management:

Goal 1. Actual results and performances are tracked against the software plans.

Goal 2. Corrective actions are taken and managed to closure when actual results and performance deviate significantly from the software plans.

Software project tracking and oversight entails tracking and reviewing software accomplishments and results against documented estimates, commitments, and plans. A documented plan for the software project is used as the basis for tracking the software activities, communicating status, and revision. Progress is primarily determined by comparing the actual software size, effort, cost, and schedule to the plan when selected software work products are completed and at selected milestones. When it is determined that the software project's plans are not being met, corrective actions are taken. These actions may include revising the software development plan to reflect the actual accomplishments and replanning.

IEEE and CMM® Software Project Tracking and Oversight

There is no one IEEE Standard that addresses software project tracking and oversight by focusing singularly on this topic. If a single standard had to be identified, IEEE Std 1058 for Software Project Management Plans could be seen to handle the majority of the CMM® Level 2 requirements for software project tracking and oversight. This standard has been covered in a previous section of this book. However, to single 1058 out as the singular standard supporting this KPA would be misleading. The goals and supporting practices are met by the common application of a number of IEEE software engineering standards. One of these supporting standards is the IEEE Standard for Software Quality Metrics, IEEE Std 1061.

IEEE Standard for Software Quality Metrics Methodology IEEE Std 1061™-1998

This standard gives a process for constructing and implementing a software quality metrics framework that can be tailor-made to meet quality requirements for a particular project and/or organization but does not mandate specific metrics for use. It defines a methodology for establishing quality requirements and identifying, implementing, analyzing, and validating the process and product software quality metrics. This methodology spans the entire software life cycle. Annex A of this standard also provides useful sample metrics validation calculations.

The goal of this standard is to provide an organization with a metrics framework that can be used to determine, define, implement, and validate quality metrics. Specifically, the use of this standard will allow an organization to:

1. Achieve quality goals
2. Establish quality requirements for a system at its outset
3. Establish acceptance criteria and standards
4. Evaluate the level of quality achieved against the established requirements
5. Detect anomalies or point to potential problems in the system
6. Predict the level of quality that will be achieved in the future
7. Monitor changes in quality when software is modified
8. Assess the ease of change to the system during product evolution
9. Validate a metric set

IEEE Std 1061 provides the step-by-step process required for implementing metrics in support of software quality measurement.

The following matrix provides a cross-reference of the SEI CMM® Level 2 Software Project Tracking and Oversight KPA to relevant supporting documentation from the IEEE Software Engineering Standards Collection. This is not meant to be an exhaustive list, rather is used to illustrate how the various KPAs are directly supported by IEEE Standards.

IEEE/CMM® Software Project Tracking and Oversight Matrix

CMM® Goals and Key Practices	Definiton	IEEE Source	Comments
G1	Actual results and performance are tracked against the software plans.		
AC1	A documented software development plan is used for tracking the software activities and communicating status.	IEEE Std 1058-1998	Derived from this IEEE standard for software project management plans.
		IEEE Std 1219-1998	Derived from this IEEE standard for software maintenance.

IEEE/CMM® Software Project Tracking and Oversight Matrix (*continued*)

CMM® Goals and Key Practices	Definiton	IEEE Source	Comments
AC1 (*cont.*)		IEEE Std 12207.1-1996	Section 6, specific information on item content guidelines.
AC5	The sizes of the software work products (or sizes of the changes to the software work products) are tracked, and corrective actions are taken as necessary. Note: This Activity simply means that a requirement is given an initial size/effort estimate and if this changes, then the schedule is adjusted or other corrective action is taken.	IEEE Std. 830-1998	Section 4.3.8, requirements should be traceable—supported if size estimates are part of the initial requirement.
		IEEE Std 1058-1998	Section 4.5, SPMP document estimates and how they were derived; it states that changes should be tracked but not how.
		IEEE Std 1219-1998	Section 4, maintenance size estimates.
		IEEE Std 1045-1992	This standard addresses the identification of the sizes of work products but is complex.
AC6	The project's software effort and costs are tracked, and corrective actions are taken as necessary.	IEEE Std 1058-1998	Section 4.5, SPMP document estimates and how they were derived.
		IEEE Std 1045-1992	Can use the examples provided in this plan.
		IEEE Std 1219-1998	Section 4, requirement for estimate, but no detail on how to derive these.
AC7	The project's critical computer resources are tracked, and corrective actions are taken as necessary.	IEEE Std 1058-1998	Section 4.5, SPMP documents require the documentation of not only critical computer resources, but also the software tools, special testing and simulation facilities, and administrative support required.
		IEEE Std 1219-1998	Section A.3.5.5, resource allocation.
AC8	The .project's software schedule is tracked, and corrective actions are taken as necessary.	IEEE Std 1058-1998	Section 4.1.1.4, schedule and budget summary.
		IEEE Std 1219-1998	Section A3.
		IEEE Std 1061-1998	Section 4.1, establish software quality requirements.
AC9	Software engineering technical activities are tracked, and corrective actions are taken as necessary.	IEEE Std 1058-1998	Section 4.6, tracking technical process.
		IEEE Std 1219-1998	Section A.3.5.6, tracking.

IEEE/CMM® Software Project Tracking and Oversight Matrix (continued)

CMM® Goals and Key Practices	Definiton	IEEE Source	Comments
AC10	The software risks associated with cost, resource, schedule, and technical aspects of the project are tracked.	IEEE Std 1058-1998	Section 4.5.3.6 of this document or Section 5 of SPMP.
		IEEE Std 1219-1998	Software maintenance. Sections 4 and A.8, risk assessment.
		IEEE Std 730-2002	Section 4.15 (Section 15 of SQAP), risk management.
AC11	Actual measurement data and replanning data for the software project are recorded.	IEEE Std 1058-1998	Section 4.5.3.6 of this document or Section 5.3.6 of SPMP.
		IEEE Std 1219-1998	Section A.3.5.6, tracking.
AC12	The software engineering group conducts periodic internal reviews to track technical progress, plans, performance, and issues against the software development plan.	IEEE Std 1058-1998	Section 4.5.3, control plan; 4.5.3.2, schedule control plan; 4.5.3.5, reporting plan.
AC13	Formal reviews to address the accomplishments and results of the software project are conducted at selected project milestones according to a documented procedure.	IEEE Std 1058-1998	Section 4.5.3.4, quality control plan identifies formal review requirement; Section 4.7, supporting process plans.
G2	Corrective actions are taken and managed to closure when actual results and performance deviate significantly from the software plans.		
AC2	The project's software development plan is revised according to a documented procedure.	IEEE Std 1058-1998	Section 4, Elements of the software project management plan; and 4.1.2, evolution of the SPMP.
		IEEE Std 730-2002	Section 4.6.2.8, managerial reviews.
AC5	The sizes of the software work products (or sizes of the changes to the software work products) are tracked, and corrective actions are taken as necessary.	IEEE Std 1058-1998	Section 5.3; control plan.
		IEEE Std 1061-1998	Software quality metrics methodology.

IEEE/CMM® Software Project Tracking and Oversight Matrix (*continued*)

CMM® Goals and Key Practices	Definiton	IEEE Source	Comments
AC6	The project's software effort and costs are tracked, and corrective actions are taken as necessary.	IEEE Std 1058-1998	Section 5.3, control plan.
		IEEE Std 1061-1998	Software quality metrics methodology.
AC7	The project's critical computer resources are tracked, and corrective actions are taken as necessary.	IEEE Std 1058-1998	Section 4.5.2.3, resource allocation.
AC8	The .project's software schedule is tracked, and corrective actions are taken as necessary.	IEEE Std 1058-1998	Section 4.5.3.2, schedule control plan.
AC9	Software engineering technical activities are tracked, and corrective actions are taken as necessary.	IEEE Std 1058-1998	Section 4.5.3.5, reporting plan.
		IEEE Std 1061-1998	Software quality metrics methodology.
AC11	Actual measurement data and replanning data for the software project are recorded.	IEEE Std 1058-1998	Section 4.5.3.6, metrics collection plan.
		IEEE Std 1061-1998	Software quality metrics methodology.
G3	Changes to software commitments are agreed to by the affected groups and individuals.		
AC3	Software project commitments and changes to commitments made to individuals and groups external to the organization are reviewed with senior management according to a documented procedure.	IEEE Std 1058-1998	Section 4.6.1, process model section. Indicates that this initial commitment information needs to be communicated, but the language is too general to indicate a requirement for communication of changes to commitments.
AC4	Approved changes to commitments that affect the software project are communicated to the members of the software engineering group and other software-related groups.	IEEE Std 1058-1998	Section 4.6.1, process model section. Indicates that this initial type of information needs to be communicated but the language is too general to indicate a requirement to communicate a change in a commitment.
CO1	A project software manager is designated to be responsible for the project's software activities and results.	IEEE Std 1058-1998	Section 4, individual responsible for plan and project; and Section 4.4, project organization.

IEEE/CMM® Software Project Tracking and Oversight Matrix (*continued*)

CMM® Goals and Key Practices	Definiton	IEEE Source	Comments
CO2	The project follows a written organizational policy for managing the software project.	IEEE Std 1058-1998	Must create policy tailored from this standard. No requirement in standard for policy statement.
AB1	A software development plan for the software project is documented and approved.	IEEE Std 1058-1998 IEEE Std 12207.1-1996	Use this standard as model for plan. Section 6, specific information item content guidelines.
AB2	The project software manager explicitly assigns responsibility for software work products and activities.	IEEE Std 1058-1998	Section 4.5.2.1, work activities.
AB3	Adequate .resources and funding are provided for tracking the software project.	IEEE Std 1058-1998	Section 4.5.1.3, resource acquisition plan; Section 4.5.2.3, resource allocation; and Section 4.5.3, control plan.
AB4	The software managers are trained in managing the technical and personnel aspects of the software project.	IEEE Std 1058-1998	Section 4.5.1.2, staffing plan; and Section 4.5.1.4, project staff training plan.
AB5	First-line software managers receive orientation in the technical aspects of the software project.	IEEE Std 1058-1998	Section 4.5.2.1, work activities; Section 4.6.1, process model; and Section 6.2, Methods, tools, and techniques.
ME1	Measurements are made and used to determine the status of the software tracking and oversight activities.	IEEE Std 1058-1998 IEEE Std 1061-1998	Section 4.5.3.6, metrics collection plan. Software quality metrics methodology.
VE1	The activities for software project tracking and oversight are reviewed with senior management on a periodic basis.	IEEE Std 1058-1998	Section 4.5.3.5, reporting plan.
VE2	The activities for software project tracking and oversight are reviewed with the project manager on both a periodic and event-driven basis.	IEEE Std 1058-1998	Section 4.5.3.5, reporting plan; and Section 4.7.5, reviews and audits plan.
VE3	The software quality assurance group reviews and/or audits the activities and work products for software project tracking and oversight and reports the results.	IEEE Std 1058-1998 IEEE Std 730-2002	Section 4.7.4, quality assurance plan. Use this standard as model for SQA plan.

Software Project Tracking and Oversight Analysis

IEEE standards can be very useful resources in support of the Software Project Tracking and Oversight KPA. Software measurement is commonly identified as a weak area during Level 2 assessments. This aspect of the Software Project Tracking and Oversight KPA could be addressed more vigilantly and, specifically, in IEEE Std 1058-1998.

The tracking of critical computer resources, work products, effort and cost are adequately handled by IEEE Std 1058; however, it is important to note that if the IEEE Standard for Software Maintenance (IEEE Std 1219) is used, support for these activities would be weak.

Please refer to the Project Planning Analysis section of this chapter (page 58) to view the recommended changes to the basic outline of the IEEE project plan suggested in Std 1058.

Example of IEEE KPA Support for Project Tracking and Oversight

CMM® AC8. The .project's software schedule is tracked, and corrective actions are taken as necessary.

IEEE Std 1058 Section 4.5.3.2 Schedule Control Plan. 4.5.3.2 Schedule control plan (Subclause 5.3.2 of the SPMP). This subclause of the SPMP specifies the control mechanisms to be used to measure the progress of work completed at the major and minor project milestones, to compare actual progress to planned progress, and to implement corrective action when actual progress does not conform to planned progress. The schedule control plan specifies the methods and tools that will be used to measure and control schedule progress. Achievement of schedule milestones should be assessed using objective criteria to measure the scope and quality of work products completed at each milestone.

The Goals of Software Project Tracking and Oversight Revisited

At the center of this KPA is a documented project plan that contains estimates for resources, effort, and cost, and an estimate-based schedule. This plan is developed after determining the commitments that have been made, the available resources, and any risk. This KPA is focused on ensuring that all of the project effectively monitors its activities and takes control actions when appropriate.

Goal 1. Actual results and performances are tracked against the software plans. This goal can effectively be met by documenting project schedule. This goal is supported by IEEE Std 1058, which specifically calls out the definition of a schedule and required schedule inputs. Using a schedule to track predictives against actuals get to the heart of this KPA.

Goal 2. Corrective actions are taken and managed to closure when actual results and performance deviate significantly from the software plans. The software development plan should not remain static, as discussed in IEEE Std 1058. All changes should be reflected as updates to the software development plan.

Goal 3. Changes to software commitments are agreed to by the affected groups and individuals. An easy way to support this goal is the addition of a signature page to the front of the project software development plan. Each signature authority signs this to indicate their acceptance of the commitments contained within.

PROJECT MONITORING AND CONTROL

The Goals of Project Monitoring and Control

The purpose of project monitoring and control is to provide an understanding of the project's progress so that appropriate corrective actions can be taken when the project's performance begins to deviate significantly from the plan. A project's documented plan forms the basis for monitoring activities, communicating status, and taking corrective action.

> SG1. Monitor Project Against Plan. Actual performance and progress of the project are monitored against the project plan.
>
> SG2. Manage Corrective Action to Closure. Corrective actions are managed to closure when the project's performance or results deviate significantly from the plan.
>
> GG2. Institutionalize a Managed Process. The process is institutionalized as a managed process.

Progress is primarily determined by comparing actual work product and task attributes, effort, cost, and schedule to the planned estimates. These comparisons occur at prescribed milestones that are identified in the project schedule or work-breakdown structure. This visibility enables a proactive response when performance deviates significantly from what is planned. A deviation is significant if, when left unresolved, it precludes the project from meeting its objectives.

IEEE and CMMI®-SW (Staged) Software Project Monitoring and Control

There is no one IEEE Standard that addresses software project monitoring and control by focusing singularly on this topic.

IEEE Standard for Software Reviews IEEE Std 1028

The IEEE 1028 standard for software reviews describes a set of review types and processes applicable throughout the software life cycle. The standard describes technical reviews, inspections, and walkthroughs. The necessary steps required for each review type are described in detail. The standard also describes the software audit process. An annex is also available that supports the process of review selection, providing a comparison of the different types of reviews.

Many organizations use informal reviews, not implementing the reviews formally as an integrated, required, part of their software process. Too often, review types are mixed and simply named "reviews" or "inspections." A walkthrough, focused on achieving consensus, is something very different from an inspection focused on formal defect finding. It is important to have a clear understanding of the desired benefit from the implementation of each type of review prior to its implementation. IEEE Std 1028 provides clear definitions and straightforward process steps. This standard describes the different review types and their specific objectives, providing a complete reference framework and common terminology. Reviewing, when applied correctly and early in the process, remains the most effective and efficient defect detection technique.[1]

---

[1] Veendendaal et al. [83]

IEEE/CMMI®-SW (Staged) Project Monitoring and Control Matrix

CMMI® Goals and Processes	Definition	IEEE Source	Comments
SG1	Monitor project against plan.		
SP 1.1	Monitor project planning parameters. Monitor the actual values of the project planning parameters against the project plan.	IEEE Std 1058-1998	Section 4.5.2.3, resource allocation; Section 4.5.3.2, schedule control plan; Section 4.5.3.5, metrics collection plan; Section 5.3, control plan.
		IEEE Std 1061-1998	Software quality metrics methodology.
SP 1.2	Monitor commitments. Monitor commitments against those identified in the project plan.	IEEE Std 1058-1998	Reporting plan, Section 4.5.3.6.
SP 1.3	Monitor project risks. Monitor risks against those identified in the project plan.	IEEE Std 1058-1998	Section 4.5.3.6 of this document or Section 5 of SPMP.
		IEEE Std 1219-1998	Software maintenance. Sections 4 and A.8, risk assessment.
		IEEE Std 730-2002	Section 4.15 (Section 15 of SQAP), risk management.
		IEEE Std 12207.1-1996	Section 6, specific information item content guidelines.
SP 1.4	Monitor data management. Monitor the management of project data against the project plan.	IEEE Std 1058-1998	Derived from this IEEE standard for software project management plans.
		IEEE Std 1219-1998	Derived from this IEEE standard for software maintenance.
SP 1.5	Monitor stakeholder involvement. Monitor stakeholder involvement against the project plan.	IEEE Std 1058-1998	Section 4.6.1, process model section indicates that this initial commitment information needs to be communicated, but the language is too general to indicate a requirement for communication of changes to commitments.
SP 1.6	Conduct progress reviews. Periodically review the project's progress, performance, and issues.	IEEE Std 1058-1998	Section 4.5.3.5, reporting plan; Section 4.7.5, reviews and audits plan.
SP 1.7	Conduct milestone reviews. Review the accomplishments and results of the project at selected project milestones.	IEEE Std 1058-1998	Section 4.5.3.5, reporting plan; Section 4.7.5, reviews and audits plan; Section 4.5.3.2, schedule control plan.
SG2	Manage corrective action to closure.		

IEEE/CMMI®-SW (Staged) Project Monitoring and Control Matrix (*continued*)

CMMI® Goals and Processes		Definition	IEEE Source	Comments
SP 2.1		Analyze issues. Collect and analyze the issues and determine the corrective actions necessary to address the issues.	IEEE Std 1058-1998	Section 4, elements of the software project management plan; 4.1.2, evolution of the SPMP; Section 4.5.3.5, reporting plan; Section 5.3, control plan.
			IEEE Std 730-2002	Section 4.6.2.8, managerial reviews.
			IEEE Std 830-1998	Section 4.3.8, requirements should be traceable—supported if size estimates are part of the initial requirement.
SP 2.2		Take corrective action. Take corrective action on identified issues.	IEEE Std 1058-1998	Section 4, elements of the software project management plan—document estimates and how they were derived, it states that changes should be tracked but not how; Section 4.1.2, evolution of the SPMP; Section 4.5.3.5, reporting plan; Section 5.3, control plan.
			IEEE Std 730-2002	Section 4.6.2.8, managerial reviews.
SP 2.3		Manage corrective action. Manage corrective actions to closure.	IEEE Std 1058-1998	Section 4, elements of the software project management plan; 4.1.2, evolution of the SPMP; Section 4.5.3.5, reporting plan; Section 5.3, control plan.
			IEEE Std 730-2002	Section 4.6.2.8, managerial reviews.
GG2		Institutionalize a managed process.		
	GP 2.1 (CO1)	Establish an organizational policy. Establish and maintain an organizational policy for planning and performing the project monitoring and control process.	IEEE Std 1058-1998	Must create policy tailored from this standard. No requirement in standard for policy statement.
	GP 2.2 (AB1)	Plan the process. Establish and maintain the plan for performing the project monitoring and control process.	IEEE Std 1058-1998	Section 4, elements of the software project management plan—document estimates and how they were derived. It states that changes should be tracked but not how. Section 4.1.2, evolution of the SPMP; Section 4.5.3.5, reporting plan; Section 5.3, control plan.
			IEEE Std 12207.1-1996	Section 6, specific information item content guidelines.

IEEE/CMMI®-SW (Staged) Project Monitoring and Control Matrix (*continued*)

CMMI® Goals and Processes	Definition	IEEE Source	Comments
GP 2.3 (AB 2)	Provide resources. Provide adequate resources for performing the project monitoring and control process, developing the work products, and providing the services of the process.	IEEE Std 1058-1998	Section 4.5.1.3, resource acquisition plan; Section 4.5.2.3, resource allocation; Section 4.5.3, control plan.
GP 2.4 (AB3)	Assign responsibility. Assign responsibility and authority for performing the process, developing the work products, and providing the services of the project monitoring and control process.	IEEE Std 1058-1998	Section 4.5.2.1, work activities.
GP 2.5 (AB 4)	Train people. Train the people performing or supporting the project monitoring and control process as needed.	IEEE Std 1058-1998	Section 4.5.1.2, staffing plan; Section 4.5.1.4, project staff training plan; Section 4.7.4, quality assurance plan.
		IEEE Std 730-2002	Use this standard as model for SQA plan.
GP 2.6 (DI1)	Manage configurations. Place designated work products of the project monitoring and control process under appropriate levels of configuration management.	IEEE Std 1058-1998	Section 4.7.1, configuration management plan—supporting plans.
		IEEE Std 828-1998	Use this standard as model for SCM plan.
GP 2.7 (DI2)	Identify and involve relevant stakeholders. Identify and involve the relevant stakeholders of the project monitoring and control process as planned.	IEEE Std 1058-1998	Section 4.5.3.4, quality control plan identifies formal review requirement; Section 4.6.1, process model section indicates that this initial type of information needs to be communicated but the language is too general to indicate a requirement to communicate a change in a commitment; Section 4.7 supporting process plans.
GP 2.8 (DI3)	Monitor and control the process. Monitor and control the project monitoring and control process against the plan for performing the process and take appropriate corrective action.	IEEE Std 1058-1998	Section 4.5.3.5, reporting plan; Section 4.7.5, reviews and audits plan; Section 4.5.3.2, schedule control plan.
GP 2.9 (VE 1)	Objectively evaluate adherence. Objectively evaluate adherence of the project monitoring and control process against its process description, standards, and procedures, and address noncompliances.	IEEE Std 1058-1998	Section 4.7.4, quality assurance plan.
		IEEE Std 730-2002	Use this standard as model for SQA plan.

IEEE/CMMI®-SW (Staged) Project Monitoring and Control Matrix (*continued*)

CMMI® Goals and Processes	Definition	IEEE Source	Comments
GP 2.10 (VE2)	Review status with higher-level management. Review the activities, status, and results of the project monitoring and control process with higher-level management and resolve issues.	IEEE Std 1058-1998	Section 4.5.3.5, reporting plan.

Project Monitoring and Control Analysis

Example of IEEE KPA Support for Project Monitoring and Control

CMMI®-SW (Staged) SP 1.7. Conduct milestone reviews. Review the accomplishments and results of the project at selected project milestones.

IEEE Std 1058 Section 4.5.3.2 Schedule Control Plan. 4.5.3.2 Schedule control plan (Subclause 5.3.2 of the SPMP). This subclause of the SPMP specifies the control mechanisms to be used to measure the progress of work completed at the major and minor project milestones, to compare actual progress to planned progress, and to implement corrective action when actual progress does not conform to planned progress. The schedule control plan specifies the methods and tools that will be used to measure and control schedule progress. Achievement of schedule milestones should be assessed using objective criteria to measure the scope and quality of work products completed at each milestone.

The Goals of Project Monitoring and Control Revisited. The CMMI®-SW (Staged) has elevated the monitoring of project commitments to the specific practice level and the effective monitoring of risk and stakeholder involvement is more strongly emphasized in the CMMI®. The project management plan becomes the basis for monitoring activities, communicating status, and taking corrective action. Progress is measured against this plan by comparing actual work product and task attributes, effort, cost, and schedule at prescribed milestones within the project schedule or work breakdown structure.

> SG1. Monitor project against plan. Actual performance and progress of the project are monitored against the project plan.
> SG2. Manage corrective action to closure. Corrective actions are managed to closure when the project's performance or results deviate significantly from the plan.
> GG2. Institutionalize a managed process. The process is institutionalized as a managed process.

The CMMI®-SW (Staged) specifically requires that the process be institutionalized as a managed process and that there be a plan for performing project monitoring and control activities. The IEEE standards listed in the previous matrix provide support for these requirements.

SOFTWARE QUALITY ASSURANCE

Quality assurance as defined by IEEE 610.12 is

> quality assurance (QA). (1) A planned and systematic pattern of all actions necessary to provide adequate confidence that an item or product conforms to established technical requirements. (2) A set of activities designed to evaluate the process by which products are developed or manufactured. [10]

Simply put, the purpose of software quality assurance (SQA) should be to specify all actions that need to be performed to check the quality of work.

A software quality assurance plan (SQAP) should take a broad view of quality and should focus on the quality of both the final product and any intermediate products. The SQAP should specify the quality assurance tasks that need to be undertaken. It should specify when in the life cycle these activities should occur and should also specify how the SQA activities are to be managed.

To summarize, a software quality assurance plan should be comprised of:

Product and process evaluations
Methods used for tracking matters requiring correction
Use of data and analyses relating to quality
Outline of the qualification procedures that will be followed
Quality interface with customers or users
Quality interface with software suppliers
Define who is responsible for what

SQA objectively evaluates performed processes, work products, and services against the established process descriptions, standards, and procedures. It identifies and documents noncompliance. All evaluation results are used to providing feedback to project staff and managers on SQA activities, ensuring that noncompliance issues are addressed

The Goals for CMM® Software Quality Assurance

This KPA provides an independent check of an organization's process. It gives management visibility into the process. SQA identifies when control actions of project tracking and oversight are not adequate and is responsible for notifying senior management.

Goal 1. SQA activities are planned.
Goal 2. Adherence of software work products and activities to the applicable standards, procedures, and requirements is verified objectively.
Goal 3. Affected groups and individuals are informed of SQA activities and results.
Goal 4. Noncompliance issues that cannot be resolved within the software project are addressed by senior management.

Software quality assurance involves reviewing and auditing software products and activities to verify that they comply with applicable procedures and standards. The results of these reviews and audits are provided to the software project manager and other appropriate managers. SQA assures that the software project establishes plans, standards, and procedures that will add value to the software project and satisfy the constraints of the project and the organization's policies.

By participating in establishing the plans, standards, and procedures, SQA helps ensure that they fit the project's needs and verifies that they will be usable for performing reviews and audits throughout the software life cycle. SQA is responsible for the review of project activities and audits software work products throughout the project life cycle. SQA provides management with visibility as to whether the software project is adhering to its established plans, standards, and procedures.

Compliance issues are first addressed within the software project and resolved there if possible. For issues not resolvable within the software project, SQA is required to escalate issues to an appropriate level of management for resolution. This KPA covers the practices for the individuals performing software quality assurance functions. The practices identifying the specific activities and work products that SQA reviews and/or audits are generally contained in the Verifying Implementation common feature of the other key SW-CMM® process areas.

The Goals for CMMI®-SW (Staged) Process and Product Quality Assurance

The Process and Product Quality Assurance KPA supports the delivery of high-quality products and services by providing the project staff and managers with visibility into the processes, and associated work products, throughout the project life cycle. Simply said, QA determines whether the project is doing what it is supposed to, based on what was originally proposed.

Objectivity in process and product quality assurance evaluations is emphasized and is critical to the successful implementation of this KPA. Traditionally, a QA group that is independent of the project provides this objectivity. Those performing QA activities for a work product should be separate from those directly involved in developing or maintaining the work product. As with the SW-CMM®, emphasis is placed on the requirement that an independent reporting channel be available so that noncompliance issues may be escalated if necessary.

> SG1. Objectively evaluate processes and work products. Adherence of the Performed process and associated work products and services to applicable process descriptions, standards, and procedures is objectively evaluated.
>
> SG2. Provide objective insight. Noncompliance issues are objectively tracked and communicated, and resolution is ensured.
>
> CG2. Institutionalize a managed process. The process is institutionalized as a managed process.

This KPA requires that quality assurance be implemented in the early phases of a project. This is so that those performing QA duties are available to participate in establishing the plans, processes, standards, and procedures that support project requirements. During this time, the specific processes, and associated work products, that will be evaluated during the project are also identified.

Supporting IEEE Software Engineering Standards

IEEE Standard for Software Quality Assurance Plans IEEE Std 730-2002.

There are distinct advantages to following, IEEE Std 730, the IEEE Software Quality Assurance Plan (SQAP). Representatives of many organizations involved in the assurance of software quality have been involved in the development of this standard. As with any IEEE software engineering standard, input was received from industry and universities, the members of the working group, and the balloting pool. The depth and breadth of experience of this collective of individuals is reflected in this document. The IEEE SQAP is directed toward the development and maintenance of critical software, irrespective of size.

The roles of quality control and quality assurance should be clarified when discussing the tasks associated with software quality. Quality control is the activities intended to find defects that exist in software work products (e.g., test, inspections). Actions in support of quality assurance are used to verify that established software development procedures and quality practices have been used as described. The recommended approaches to good quality assurance practices are referred to by IEEE Std 730.1-1998. This plan basically provides a detailed description of the documentation format and content required for an SQAP.

The following matrix provides a cross-reference of the SEI CMM® Level 2 Software Quality Assurance KPA to relevant supporting documentation from the IEEE Software Engineering Standards Collection. This is not meant to be an exhaustive list, but rather is used to illustrate how the various KPAs are directly supported by IEEE standards.

IEEE Guide for Software Quality Assurance Planning IEEE Std 730.1-1998.

This guide was developed as a supplement to the IEEE Std 730-1989. It is still available from IEEE, but is not included in the current Software Engineering Standards Collection. This standard is not being maintained (i.e., not being revised every 5 years) because sections of this guide were incorporated into the IEEE Software Quality Assurance Std 730, when it was revised in 2002. This guide is listed here as a resource because it contains detailed information in support of the creation, maintenance, evaluation, and modification of a software quality assurance plan. This guide focuses specifically on the SQAP and not the practice of SQA as a whole. This guide may be used as a companion document to IEEE Std 730.

IEEE/CMM® Software Quality Assurance Matrix

CMM® Goals and Key Practices	Definiton	IEEE Source	Comments
G1	Software quality assurance activities are planned.		
AC1	A SQA plan is prepared for the software project according to a documented procedure.	IEEE Std 730-2002, IEEE Std 730.1-1995	These standards may be used as a template for the organization SQA Plan.
		IEEE Std 12207.1-1996	Section 6, specific information item content guidelines.

IEEE/CMM® Software Quality Assurance Matrix (continued)			
CMM® Goals and Key Practices	Definiton	IEEE Source	Comments
AC2	The SQA group's activities are performed in accordance with the SQA plan.	IEEE Std 730-2002	Section 4.3.1, organization; Section 4.3.2, tasks; and Section 4.3.3 roles and responsibilities.
		IEEE Std 730.1-1995	Section 3.3.2, tasks.
G2	Adherence of software products and activities to the applicable standards, procedures, and requirements is verified objectively.		
AC2	The SQA group's activities are performed in accordance with the SQA plan.	IEEE Std 730-2002	Section 4.3.1, organization; Section 4.3.2, tasks; and Section 4.3.3, responsibilities.
		IEEE Std 730.1-1995	Section 3.3.2, tasks.
AC3	The SQA group participates in the preparation and review of the project's software development plan, standards, and procedures.	IEEE Std 730-2002	Section 4.4.2, minimum documentation requirements. Note: The standard states clearly which document reviews are required.
		IEEE Std 730.1-1995	Section 3.4, documentation.
AC4	The SQA group reviews the software engineering activities to verify compliance.	IEEE Std 730-2002	Section 4.3.1, organization; Section 4.3.2; tasks; and Section 4.3.3, responsibilities.
		IEEE Std 730.1-1995	Section 3.3.2, tasks. Note: The standard provides detailed information on the involvement of SQA in software engineering activities.
AC5	The SQA group audits designated software work products to verify compliance.	IEEE Std 730-2002	Section 4.6 (Section 6 of SQAP), reviews and audits.
		IEEE Std 730.1-1995	Section 3.6, reviews and audits.
G3	Affected groups and individuals are informed of software quality assurance activities and results.		
AC6	The SQA group periodically reports the results of its activities to the software engineering group.	IEEE Std 730-2002	Section 4.5.1, Purpose. Add providing feedback to the software engineering group to purpose of practices, conventions and metrics section.
			Section 4.8, problem reporting and corrective action. Need to specify all reporting, not just problem reporting.

IEEE/CMM® Software Quality Assurance Matrix (*continued*)

CMM® Goals and Key Practices	Definiton	IEEE Source	Comments
AC6 (*cont.*)		IEEE Std 730.1-1995	Section 3.8, problem reporting and corrective action (Section 8 of SQAP).
AC7	Deviations identified in the software activities and software work products are documented and handled according to a documented procedure.	IEEE Std 730-2002	Section 4.8, problem reporting and corrective action.
		IEEE Std 730.1-1995	Section 3.8, problem reporting and corrective action (Section 8 of SQAP).
AC8	The SQA group conducts periodic reviews of its activities and findings with the customer's SQA personnel, as appropriate.	IEEE Std 730.1-1995	Section 3.5.2.5, maintenance phase.
G4	Noncompliance issues that cannot be resolved within the software project are addressed by senior management.		
AC7	Deviations identified in the software activities and software work products are documented and handled according to a documented procedure.	IEEE Std 730-2002	Section 4.8, problem reporting and corrective action.
		IEEE Std 730.1-1995	Section 3.8, problem reporting and corrective action (Section 8 of SQAP).
CO1	The project follows a written organizational policy for implementing software quality assurance (SQA).	IEEE Std 730-2002 730.1-1995	No requirement in standard for policy statement.
AB1	A group that is responsible for coordinating and implementing SQA for the project (i.e., the SQA group) exists.	IEEE Std 730.1-1995	Section 4.3, planning for the implementation of the SQAP; and Section 4.3.1, resources.
		IEEE Std 730-2002	Section 4.3 (Section 3 of SQAP), management, i.e., tasks and responsibilities.
AB2	Adequate resources and funding are provided for performing the SQA activities.	IEEE Std 730.1-1995	Section 4.3, planning for the Implementation of the SQAP; and Section 4.3.1, resources.
AB3	Members of the SQA group are trained to perform their SQA activities.	IEEE Std 730-2002	Section 4.14 (Section 14 of SQAP), training.
		IEEE Std 730.1-1995	Section 3.14, training.
		IEEE Std 1058-1998	Section 4.5.1.4, project staff training plan; section calls for a training plan, up to the PM to include SQA activity training.

IEEE/CMM® Software Quality Assurance Matrix (*continued*)

CMM® Goals and Key Practices	Definiton	IEEE Source	Comments
AB4	The members of the software project receive orientation on the role, responsibilities, authority, and value of the SQA group.	IEEE Std 730-2002	Information required on training should include the orientation of members of the software project.
		IEEE Std 730.1-1995	
		IEEE Std 1058-1998	Section 4.5.1.4, project staff training plan; section calls for a training plan, up to the PM to include SQA activity training.
ME1	Measurements are made and used to determine the cost and schedule status of the SQA activities.	IEEE Std 730-2002	No requirement found in standard.
		IEEE Std 730.1-1995	
VE1	The SQA activities are reviewed with senior management on a periodic basis.	IEEE Std 730-2002	Section 4.6.2.8, managerial reviews.
		IEEE Std 730.1-1995	Section 4.2, acceptance by management; and Section 3.6, reviews and audits.
VE2	The SQA activities are reviewed with the project manager on both a periodic and event-driven basis.	IEEE Std 1058-1998	Section 4.5.3.4, quality control plan (Subclause 5.3.4 of the SPMP).
		IEEE Std 730-2002	Section 4.3, management; Section 4.6, reviews and audits.
VE3	Experts independent of the SQA group periodically review the activities and software work products of the project's SQA group.	IEEE Std 730-2002	Section 4.3.1, organization (Section 3 of the SQAP).
	Note: This is not explicitly called out in Std 730-2002, but if the explanation that is found in Std 730.1-1995 is used as a guide, then this criterion is met.	IEEE Std 730.1-1995	Section 3.3.1, organization.

IEEE/CMMI®-SW (Staged) Process & Product Quality Assurance Matrix

CMMI® Goals and Processes	Definition	IEEE Source	Comments
SG1	Objectively evaluate processes and work products.		
SP 1.1	Objectively evaluate processes. Objectively evaluate the designated performed processes against the applicable process descriptions, standards, and procedures.	IEEE Std 730-2002	Section 4.6, software reviews; type of reviews and minimum requirements
SP 1.2	Objectively evaluate work products and services. Objectively evaluate the designated work products and services against the applicable process descriptions, standards, and procedures.	IEEE Std 1061-1998	Section 4, software metrics methodology; guidance on establishing quality requirements, quality metrics, implementation, results analysis, and validation.
SG2	Provide objective insight.		
SP 2.1	Communicate and ensure resolution of noncompliance issues. Communicate quality issues and ensure resolution of noncompliance issues with the staff and managers.	IEEE Std 730-2002	Section 4.8, problem reporting and corrective action.
		IEEE Std 730.1-1995	Section 3.8, problem reporting and corrective action (Section 8 of SQAP).
SP 2.2	Establish records. Establish and maintain records of the quality assurance activities.	IEEE Std 730-2002	Section 4.8, problem reporting and corrective action.
			Section 3.8, problem reporting and corrective action (Section 8 of SQAP).
GG2	Institutionalize a managed process.		
GP 2.1 (CO1)	Establish an organizational policy. Establish and maintain an organizational policy for planning and performing the process and product quality assurance process.	IEEE Std 730-2002	Not specifically called out in standard; however, Section 4.3.1 talks to the documentation of the organizational structure in support of SQA activities.
GP 2.2 (AB1)	Plan the process. Establish and maintain the plan for performing the process and product quality assurance process.	IEEE Std 730-2002	Section 4.3.2, tasks.
		IEEE Std 12207.1-1996	Section 6, specific information item content guidelines.
GP 2.3 (AB 2)	Provide resources. Provide adequate resources for performing the process and product quality assurance process, developing the work products, and providing the services of the process.	IEEE Std 730-2002	Section 4.3.3, roles and responsibilities; and Section 4.3.4, quality assurance estimated resources.
		IEEE Std 730.1-1995	Section 4.3, planning for the implementation of the SQAP; and Section 4.3.1, resources.

IEEE/CMMI®-SW (Staged) Process & Product Quality Assurance Matrix (continued)			
CMMI® Goals and Processes	Definition	IEEE Source	Comments
GP 2.4 (AB3)	Assign responsibility. Assign responsibility and authority for performing the process, developing the work products, and providing the services of the process and product quality assurance process.	IEEE Std 730.1-1995	Section 4.3, planning for the implementation of the SQAP; and Section 4.3.1, resources.
		IEEE Std 730-2002	Section 4.3 (Section 3 of SQAP), management, i.e., tasks and responsibilities.
GP 2.5 (AB4)	Train people. Train the people performing or supporting the process and product quality assurance process as needed.	IEEE Std 730-2002	Section 4.14 (Section 14 of SQAP), training.
		IEEE Std 730.1-1995	Section 3.14, training.
		IEEE Std 1058-1998	Section 4.5.1.4, project staff training plan; section calls for a training plan, up to the PM to include product quality assurance training.
GP 2.6 (DI1)	Manage configurations. Place designated work products of the process and product quality assurance process under appropriate levels of configuration management.	IEEE Std 730-2002	Section 4.4.2.6, software configuration management plan.
GP 2.7 (DI2)	Identify and involve relevant stakeholders. Identify and involve the relevant stakeholders of the process and product quality assurance process as planned.	IEEE Std 730-2002	Section 4.3, management. Are not called "relevant stakeholders" but call in standard to identify organization, tasking, roles, and responsibilities.
GP 2.8 (DI3)	Monitor and control the process. Monitor and control the process and product quality assurance process against the plan for performing the process and take appropriate corrective action.	IEEE Std 730-2002	Section 4.6, software reviews.
GP 2.9 (VE 1)	Objectively evaluate adherence. Objectively evaluate adherence of the process and product quality assurance process against its process description, standards, and procedures, and address noncompliance.	IEEE Std 720-2002	This is essentially the QA of the SQA process, not called out in plan.
GP 2.10 (VE2)	Review status with higher-level management. Review the activities, status, and results of the process and product quality assurance process with higher-level management and resolve issues.	IEEE Std 720-2002	Section 4.6.2.8, managerial reviews.

Software Quality Assurance Analysis

The Software Engineering Institute's CMM® states that, "The purpose of software quality assurance is to provide management with appropriate visibility into the process being used by the software project and of the products being built." [48] The CMMI®-SW (Staged) defines quality assurance as "The purpose of process and product quality assurance is to provide staff and management with objective insight into processes and associated work products." [57] Both of these definitions provide complementary summaries of SQA. However, when looking at IEEE Std 730, it becomes readily apparent that feedback mechanisms and SQA activity review need to be more directly addressed.

Figure 5-4 offers a suggested modification to the outline proposed by IEEE Std 730. In order to maximize quality, improved processes are applied that reduce the number of defects introduced during software development.

Example of IEEE KPA Support for Software Quality Assurance

CMM® AC5. The SQA group audits designated software work products to verify compliance.

CMMI®-SW (Staged) SP 1.1. Objectively evaluate processes. Objectively evaluate the designated performed processes against the applicable process descriptions, standards, and procedures.

IEEE Std 730 Section 4.6 Software Reviews. This section:

a) Defines the software reviews to be conducted. They may include managerial reviews, acquirer-supplier reviews, technical reviews, inspections, walk-throughs, and audits.

b) Lists the schedule for software reviews as they relate to the software project's schedule.

c) States how the software reviews shall be accomplished.

d) States what further actions are required and how they are to be implemented and verified.

Section 4.6 goes on to further list and describe a prescribed minimum set of audits and reviews. These are simply listed in Table 5-3. Complete supporting detailed descriptions are provided in IEEE Std 730.

The Goals of Software Quality Assurance Revisited

SW-CMM® Goals for Software Quality Assurance. This KPA provides an independent check of an organizations process. It gives management visibility into the process. SQA identifies when control actions of project tracking and oversight are not adequate and is responsible for notifying senior management.

Goal 1. SQA activities are planned. The SQA plan may be distinct from the software project management plan (SPMP), or it may be integrated into the SPMP. Regardless of where SQA activities are documented, IEEE Std 730 supports the documentation of SQP group funding and resources, the activities and work products to be

> 1. Purpose
> 2. Reference documents
> 3. Management
>   a. Organization
>   b. Tasks
>   c. Responsibilities
> 4. Documentation
>   a. Minimum Documentation Requirements
>   i. Software Requirements Specification (SRS)
>   ii. Software Design Description (SDD)
>   iii. Software Verification and Validation Plan (SVVP)
>   iv. Software Verification and Validation Report (SVVR)
>   v. User Documentation
>   vi. Software Configuration Management Plan (SCMP)
>   **b. Feedback Mechanisms**
> 5. Standards, practices, conventions, and metrics
>   a. Content
> 6. Reviews and audits
>   a. Minimum Requirements
>   i. Software Requirements Review (SRR)
>   ii. Preliminary Design Review (PDR)
>   iii. Critical Design Review (CDR)
>   iv. Software Verification and Validation Plan Review (SVVPR)
>   v. Functional Audit
>   vi. Physical Audit
>   vii. In-process Audits
>   viii. Managerial Reviews
>   ix. Software Configuration Management Plan Review (SCMPR)
>   x. Post-Mortem Review
> 7. Test
> 8. Problem Reporting and Corrective Action
>   a. Problem Reporting
>   b. Corrective Action
>   **c. Feedback Mechanisms**
> **9. SQA Activity Review**
>   **a. Escalation Procedures**
>   **b. Metrics and Measurement**
>   **c. Tools, Techniques, and Methodologies**
>   **d. Process**
> 10. Code Control
> 11. Media Control
> 12. Supplier Control
> 13. Records collection, maintenance, and retention
> 14. Training
> 15. Risk Management

Figure 5-4 Example SQAP—Based on IEEE Std 730/730.1.

evaluated, and the standards and the procedures by which reviews are performed. The escalation procedures are not directly addressed by IEEE Std 730.

Goal 2. Adherence of software work products and activities to the applicable standards, procedures, and requirements is verified objectively. To effectively meet this goal, both IEEE Std 1058 for Software Project Management Plans and IEEE Std 730 for Software Quality Assurance Plans should be referenced. It is critical that the

Table 5-3 IEEE Std 730 minimum set of software reviews

Software Specification Review
Architectural Design Review
Detailed Design Review
Verification and Validation Plan Review
Functional Audit
Physical Audit
In-Process Audit
Managerial Reviews
Software Configuration Management Plan Review
Post Implementation Review

fact that the SQA group participation in the preparation and review of the SPMP is captured in either the SPMP or the SQAP. The inclusion of the plan review, work product, and audits in the SPMP project schedule would suffice. The types of reviews and audits should be defined in the associated SQAP.

Goal 3. Affected groups and individuals are informed of SQA activities and results. Again, a combination of both IEEE Std 1058 for Software Project Management Plans and IEEE Std 730 for Software Quality Assurance Plans should be referenced to adequately address this goal. The SPMP should identify reporting hierarchy and deliverable reports. The SQAP should identify the activities and how the results will be presented.

Goal 4. Noncompliance issues that cannot be resolved within the software project are addressed by senior management. The escalations of issues of nonconformance are not directly addressed by IEEE Std 730 for Software Quality Assurance Plans. This is reflected in the modified suggested table of contents provided in Figure 5-4.

CMMI®-SW (Staged) Goals for Process and Product Quality Assurance Revisited. Records collection, maintenance, and retention are directly addressed by IEEE Std 730. Emphasis is placed on these activities as described by the CMMI®-SW (Staged), SP 2.2, Establish and maintain records of the quality assurance activities. These activities were implied by the SW-CMM® but not directly addressed in the manner that they are in the CMMI®.

SG1. Objectively Evaluate Processes and Work Products. Adherence of the Performed process and associated work products and services to applicable process descriptions, standards, and procedures is objectively evaluated.

SG2. Provide Objective Insight. Noncompliance issues are objectively tracked and communicated, and resolution is ensured.

IEEE Standards 730/730.1 can be used in conjunction with the IEEE Standard for Software Project Management Plans to develop processes and the supporting documentation required to meet the SG1 and SG2 goals for CMMI®-SW (Staged) Process and Product Quality Assurance. These standards provide the details required to support SQA activities, including process development, plan documentation, and process and plan maintenance, evaluation, and modification.

CG2. Institutionalize a Managed Process. The process is institutionalized as a managed process.

The CMM® did not fully support the requirement to establish and maintain the SQAP. The CMMI® requires that the plan for performing the process and product quality assurance process. This is different from just requiring the development of an SQAP. This CMMI® requirement states that the process used to define SQA at the project level must be defined. There must be a project level SQAP, but also some type of organizational plan describing how this plan should be developed. IEEE Std 730 and 730.1 can be used to support both planning activities.

SOFTWARE CONFIGURATION MANAGEMENT

Software configuration management (SCM) is the identification, documentation, and management of hardware or software that has been deemed critical to the success of a particular software effort. SCM provides the methods and tools used to identify and control the software throughout its development and use. Typically all source code is placed under configuration control. Mature SCM practices ensure the validity of software baselines at any given point in time, the control of baseline changes, and tracking and reporting of baseline changes. To plainly describe it, work products placed under configuration management should include products that will be delivered to the customer, any internal work products, and all tools used in creating and describing these work products.

A configuration management plan should be comprised of:

Information regarding a numbering scheme for software and documents
Identification of controlled libraries and what they are used for
Tools unique to configuration management
Identify when items go under control
Specify how controlled items get changed
How change information is disseminated
Who is responsible for what

A good software configuration management plan should document what SCM activities are to be done, how they are to be done, who is responsible for doing specific activities, when they are to happen, and what resources are required. It should also be applicable to any type and size of software development effort and be able to address SCM activities over any portion of a product's life cycle.

The Goals for CMM® Software Configuration Management

This goal is the most straightforward of the Level 2 KPAs. This goal involves the control of any specific version of any organization's work products. Simply put, it means making sure that everyone has the correct version.

Goal 1. SCM activities are planned.
Goal 2. Selected software work products are identified, controlled, and available.

Goal 3. Changes to identified software work products are controlled.

Goal 4. Affected groups and individuals are informed of the status and content of software baselines.

Software configuration management addresses the identification and configuration of the software at given points in the software life cycle. SCM systematically controls changes to the configuration, and maintains the integrity and traceability of the configuration throughout the software process. Work products placed under software configuration management include the software products that are delivered and the items that are identified with or required to create these software products.

This KPA requires that a software baseline library be established. This library must contain the software baselines as they are developed. Changes to baselines and the release of software products built from the software baseline library are systematically controlled via SCM change control and configuration auditing functions. This KPA covers the practices for performing the software configuration management function. The practices identifying specific configuration items or units are contained in the key process areas that describe the development and maintenance of each configuration item/unit.

When beginning to implement process improvement practices, this KPA is often the easiest one to begin with. Developers often see immediate benefit from the implementation of software configuration management processes and procedures. This perception of direct benefit can often be leveraged into the other CMM® Level 2 KPAs.

The Goals for CMMI®-SW (Staged) Configuration Management

As described in the CMMI®-SW (Staged), "The purpose of configuration management is to establish and maintain the integrity of work products using configuration identification, configuration control, configuration status accounting, and configuration audits." [65] This KPA focuses on the expectations surrounding what is required to effectively establish a software product baseline, the control of the product baseline, and the activities required to ensure the integrity of the baseline

SG 1. Establish Baselines. Baselines of identified work products are established. This specific goal requires that a configuration management system be established, the identification of all configuration items, and that all release baselines must be established. The CMMI®-SW identifies a baseline as the set of requirements, design, source code files and the associated executable code, build files, and user documentation that have been assigned a unique identifier. Release of a baseline requires retrieval of source code files from the configuration management system.[2]

SG 2. Track and Control Changes. Changes to the work products under configuration management are tracked and controlled. It does no good to establish baselines unless these baselines are effectively maintained. This specific goal (SG) serves to maintain baselines once they are established. Baselines are not static. For example, changes to product baselines occur through requirements implementation; phase 1 of a project may be complete so the product is baselined; phase 2 begins, requiring the development of a

---

[2] A baseline that is delivered to a customer is typically called a "release," whereas a baseline for an internal use is typically called a "build."

new product baseline. Baseline changes may also be due to customer requests or changes in requirements. To ensure effective baseline management, all changes to items that collectively comprise the product baseline must be tracked.

SG 3 Establish Integrity. Integrity of baselines is established and maintained. The purpose of this SG is to ensure that each product baseline can be identified and recovered if necessary. All items included in a product baseline must be described. This description must include a complete revision history describing any changes and associated change rationale. Configuration audits are also required in support of baseline maintenance once the baseline is established.

CG2 Institutionalize a Managed Process. The process is institutionalized as a managed process. The CMMI®-SW (Staged) requires that the processes used in support of configuration management planning and project level process development be institutionalized. It requires that individuals be assigned specific responsibilities and that these individuals be granted authority to accomplish the goals of this KPA. The importance of adequate training in support of configuration management activities is also emphasized.

Supporting IEEE Software Engineering Standards

IEEE Standard for Software Configuration Management Plans IEEE Std 828-1998. There are distinct advantages to following the IEEE Software Configuration Management Plan (IEEE Std 828) standard (see Table 5-4). First, it is a standard drawn up by representatives of many organizations involved in software configuration management. Input is received from industry and universities, the members of the working group, and reviewing teams. This collection of individuals represents many years experience determining appropriate and reasonable software configuration management. Second, the IEEE SCMP is directed toward the development and maintenance of all types of software (i.e., critical, noncritical, and software in maintenance), irrespective of size.

Table 5-4 Overview of IEEE Std 828-1998

Class of Information	Description	IEEE Std 828-1998 Reference
Introduction	Describes the plan's purpose, scope of application, key terms, and references	4.1
SCM management	(Who?) Identifies the responsibilities and authorities for accomplishing the planned activities	4.2
SCM activities	(What?) Identifies all activities to be performed in applying to the project	4.3
SCM schedules	(When?) Identifies the required coordination of SCM activities with other activities in the project	4.4
SCM resources	(How?) Identifies tools and physical and human resources required for execution of the plan	4.5
SCM plan maintenance	Identifies how the plan will be kept current while in effect	4.6

Table 5-5 Cross-reference to IEEE Std 1042-1987 [12]

IEEE Std 828-1998	IEEE Std 1042-1987
Overview	SCM Disciplines in Software Management
The Software Configuration Management Plan	Software Configuration Management Plans
Introduction	Introduction
SCM Management	Management
SCM Activities	SCM Activities
Configuration Identification	Configuration Identification
Configuration Control	Configuration Control
Configuration Status Accounting	Configuration Status Accounting
Configuration Audits and Reviews	Configuration Audits and Reviews
Interface Control	Interface Control
Subcontractor Vendor Control	Supplier Control
SCM Schedules	SCM Plan Implementation
SCM Resources	Tools, Techniques, and Methodologies
SCM Plan Maintenance	
Standard Tailoring	
Standard Conformance	

As described by IEEE Std 828-1998;

SCM constitutes good engineering practice for all software projects, whether phased development, rapid prototyping, or ongoing maintenance. It enhances the reliability and quality of software by providing a structure for identifying and controlling documentation, code, interfaces, and databases to support all life cycle phases supporting a chosen development/maintenance methodology that support the requirements, standards, policies, organization, and management philosophy producing management and product information concerning the status of baselines, change control, tests, releases, audits, etc.

Recommended approaches to good SCM practices in support of this standard are referred to in the IEEE Guide to Software Configuration Management, also known as IEEE Std 1042-1987 (see Table 5-5). This guide is still available from IEEE Press; however, it is not included in the current standards collection, as it is no longer maintained for publication. The plan does have value, basically providing examples of appropriately applied SCM.

The following matrix provides a cross-reference of the SEI CMM® Level 2 Software Configuration Management KPA to relevant supporting documentation from the IEEE Software Engineering Standards Collection. This is not meant to be an exhaustive list; rather, it is used to illustrate how the various KPAs are directly supported by IEEE standards.

IEEE/CMM® Software Configuration Management Matrix

CMM® Goals and Key Practices	Definiton	IEEE Source	Comments
G1	Software configuration management activities are planned.		
AC1	A SCM plan is prepared for each software project according to a documented procedure.	IEEE Std 828-1998	This plan used as a template for the organization SCM Plan.
AC2	A documented and approved SCM plan is used as the basis for performing the SCM activities.	IEEE Std 828-1998	Section 4.2; organization; Sections 4.2.1 and 4.2.2, responsibilities; Section 4.2.3 applicable policies, directives, and procedures.
		IEEE 12207.1-1996	Section 6, specific information item content guidelines.
G2	Selected software work products are identified, controlled, and available.		
AC3	A configuration management library system is established as a repository for the software baselines.	IEEE Std 828-1998	Section 4.3, SCM activities. Specifically 4.3.1.2, naming configuration items.
AC4	The software work products to be placed under configuration management are identified.	IEEE Std 828-1998	Section 4.3.1, identifying configuration items.
AC7	Products from the software baseline library are created and their release is controlled according to a documented procedure.	IEEE Std 828-1998	Section 4.3.2, configuration control.
G3	Changes to identified software work products are controlled.		
AC5	Change requests and problem reports for all configuration items/units are initiated, recorded, reviewed, approved, and tracked according to a documented procedure.	IEEE Std 828-1998	Section 4.3.3, configuration status accounting.
AC6	Changes to baselines are controlled according to a documented procedure.	IEEE Std 828-1998	Section 4.3.2.4, implementing changes.
G4	Affected groups and individuals are informed of the status and content of software baselines.		
AC8	The status of configuration items/units is recorded according to a documented procedure.	IEEE Std 828-1998	Section 4.3.3, configuration status accounting.

IEEE/CMM® Software Configuration Management Matrix (*continued*)

CMM® Goals and Key Practices	Definiton	IEEE Source	Comments
AC9	Standard reports documenting the SCM activities and the contents of the software baseline are developed and made available to affected groups and individuals.	IEEE Std 828-1998	Section 4.3.3, configuration status accounting.
AC10	Software baseline audits are conducted according to a documented procedure.	IEEE Std 828-1998	Section 4.3.4, configuration audits and reviews.
CO1	The project follows a written organizational policy for implementing software configuration management (SCM).	IEEE Std 828-1998	Section 2.3, applicable policies, directives, and procedures.
AB1	A board having the authority for managing the project's software baselines (i.e., a software configuration control board or SCCB) exists or is established.	IEEE Std 828-1998 IEEE Std 828-1998	Section 4.3.2.3, approving or disapproving changes. Section 4.2.2, SCM responsibilities. Requires the definition of a review board if one exists.
AB2	A group that is responsible for coordinating and implementing SCM for the project (i.e., the SCM group) exists.	IEEE Std 828-1998	Section 4.2.2, SCM responsibilities.
AB3	Adequate resources and funding are provided for performing the SCM activities.	IEEE Std 1058-1998	Section 4.7, supporting process plans.
AB4	Members of the SCM group are trained in the objectives, procedures, and methods for performing their SCM activities.	IEEE Std 828-1998 IEEE Std 1058-1998	Section 4.5, SCM resources. Section 4.5.1.4, project staff training plan; section calls for a training plan, up to the PM to include SCM activity training.
AB5	Members of the software engineering group and other software-related groups are trained to perform their SCM activities.	IEEE Std 828-1998 IEEE Std 1058-1998	Section 4.5, SCM resources. Section 4.5.1.4, project staff training plan; section calls for a training plan, up to the PM to include SCM activity training.
ME1	Measurements are made and used to determine the status of the SCM activities.	IEEE Std 828-1998	Section 4.3.4, configuration audits and reviews.

IEEE/CMM® Software Configuration Management Matrix (continued)

CMM® Goals and Key Practices	Definiton	IEEE Source	Comments
VE1	The SCM activities are reviewed with senior management on a periodic basis.	IEEE Std 828-1998	Section 4.3.4, configuration audits and reviews.
VE2	The SCM activities are reviewed with the project manager on both a periodic and event-driven basis.	IEEE Std 828-1998	Section 4.3.4, configuration audits and reviews.
		IEEE Std 1058-1998	Section 7.1, configuration management plan.
VE3	The SCM group periodically audits software baselines to verify that they conform to the documentation that defines them.	IEEE Std 828-1998	Section 4.3.4, configuration audits and reviews. Not specifically called out.
		IEEE Std 730-2002	Section 4.6.2.9, software configuration management plan review.
VE4	The software quality assurance group reviews and/or audits the activities and work products for SCM and reports the results.	IEEE Std 828-1998	Section 4.3.4, configuration audits and reviews.
		IEEE Std 730-2002	Section 4.4.2.6, software configuration management plan.

IEEE/CMMI®-SW (Staged) Configuration Management Matrix

CMMI® Goals and Processes	Definition	IEEE Source	Comments
SG1	Establish baselines.		
SP 1.1	Identify configuration iItems. Identify the configuration items, components, and related work products that will be placed under configuration management.	IEEE Std 828-1998	Section 4.3.1.1, identifying configuration items; Section 4.3.1.2, naming configuration items; Section 4.3.2, configuration control.
SP 1.2	Establish a configuration management system. Establish and maintain a configuration management and change management system for controlling work products.	IEEE Std 828-1998	Section 4.3.2, configuration control.
SP 1.3	Create or release baselines. Create or release baselines for internal use and for delivery to the customer.	IEEE Std 828-1998	Section 4.3.2.4, implementing changes. Does not directly address the control of release and delivery of software products or documentation. Must specify the maintenance of master copies of released code and documentation.

IEEE/CMMI®-SW (Staged) Configuration Management Matrix

CMMI® Goals and Processes	Definition	IEEE Source	Comments
SG2	Track and control changes.		
SP 2.1	Track change requests. Track change requests for the configuration items.	IEEE Std 828-1998	Section 4.3.3, configuration status accounting.
SP 2.2	Control configuration items. Control changes to the configuration items.	IEEE Std 828-1998	Section 4.3.3, configuration status accounting.
SG3	Establish integrity.		
SP 3.1	Establish configuration management records. Establish and maintain records describing configuration items.	IEEE Std 828-1998	Section 4.3.3, configuration status accounting.
		IEEE Std 12207.1-1996	Section 6, specific information item content guidelines.
SP 3.2	Perform configuration audits. Perform configuration audits to maintain integrity of the configuration baselines.	IEEE Std 828-1998	Section 4.3.4, configuration audits and reviews. Not specifically called out.
		IEEE Std 730-2002	Section 4.6.2.9, software configuration management plan review.
GG2	Institutionalize a managed process.		
GP 2.1 (CO1)	Establish an organizational policy. Establish and maintain an organizational policy for planning and performing the configuration management process.	IEEE Std 828-1998	Section 2.3, applicable policies, directives, and procedures.
GP 2.2 (AB1)	Plan the process. Establish and maintain the plan for performing the configuration management process.	IEEE Std 828-1998	Section 4.2, organization; Sections 4.2.1 and 4.2.2, responsibilities; and Section 4.2.3, applicable policies, directives, and procedures.
		IEEE 12207.1-1996	Section 6, specific information item content guidelines.
GP 2.3 (AB 2)	Provide resources. Provide adequate resources for performing the configuration management process, developing the work products, and providing the services of the process.	IEEE Std 1058-1998	Section 4.7, Supporting process plans.
GP 2.4 (AB3)	Assign responsibility. Assign responsibility and authority for performing the process, developing the work products, and providing the services of the configuration management process.	IEEE Std 828-1998	Section 4.3.2.3, approving or disapproving changes.
		IEEE Std 828-1998	Section 4.2.2, SCM responsibilities. Requires the definition of a review board if one exists

IEEE/CMMI®-SW (Staged) Configuration Management Matrix

CMMI® Goals and Processes	Definition	IEEE Source	Comments
GP 2.5 (AB4)	Train people. Train the people performing or supporting the configuration management process as needed.	IEEE Std 828-1998	Section 4.5, SCM resources.
		IEEE Std 1058-1998	Section 4.5.1.4. project staff training plan; section calls for a training plan, up to the PM to include SCM activity training.
GP 2.6 (DI1)	Manage configurations. Place designated work products of the configuration management process under appropriate levels of configuration management.	IEEE Std 828-1998	Section 4.3.1.1, identifying configuration items; Section 4.3.1.2, naming configuration items; Section 4.3.2, configuration control.
GP 2.7 (DI2)	Identify and involve relevant stakeholders. Identify and involve the relevant stakeholders of the configuration management process as planned.	IEEE Std 828-1998	Section 4.3.2.3, approving or disapproving changes.
		IEEE Std 828-1998	Section 4.2.2, SCM responsibilities. Requires the definition of a review board if one exists
GP 2.8 (DI3)	Monitor and control the process. Monitor and control the configuration management process against the plan for performing the process and take appropriate corrective action.	IEEE Std 730-2002	Section 4.4.2.6, SQA, software configuration management plan.
GP 2.9 (VE 1)	Objectively evaluate adherence. Objectively evaluate adherence of the configuration management process against its process description, standards, and procedures, and address non-compliance.	IEEE Std 730-2002	Section 4.4.2.6, SQA, software configuration management plan.
GP 2.10 (VE2)	Review status with higher-level management. Review the activities, status, and results of the configuration management process with higher-level management and resolve issues.	IEEE Std 828-1998	Section 4.3.4, configuration audits and reviews.

Software Configuration Management Analysis

Standards provide written expectations for the processes to be used and the deliverables to be produced by software developers. IEEE Std 828 may be used to highlight the specific areas of configuration management that should be considered for process development and documentation. This standard from the IEEE Software Engineering Standards Collection directly supports most of the CMM® Level 2 SCM KPA requirements and can be used with no modification. It is important to note that this plan should be used in conjunction with the SQA plan, which would apply to items like the types of review to be performed on the SCM plan (see Figure 5-5), and the SPMP, which would address other supporting plans.

Example of IEEE KPA Support for Software Configuration Management

CMM® AC4. The software work products to be placed under configuration management are identified.

```
1.0 INTRODUCTION
    1.1 Purpose
    1.2 Scope
    1.3 Definitions/Acronyms
    1.4 References
    1.5 Tailoring
2.0 SOFTWARE CONFIGURATION MANAGEMENT
    2.1 SCM organization
    2.2 SCM responsibilities
    2.3 Relationship of CM to the software process life cycle
        2.3.1 Interfaces to other organizations on the project
        2.3.2 Other project organizations CM responsibilities
3.0 SOFTWARE CONFIGURATION MANAGEMENT ACTIVITIES
    3.1 Configuration Identification
        3.1.1 Specification Identification
        3.2.2 Change Control Form Identification
        3.2.3 Project Baselines
        3.2.4 Library
    3.2 Configuration Control
        3.2.1 Procedures for changing baselines
        3.2.2 Procedures for processing change requests and approvals
        3.2.3 Organizations assigned responsibilities for change control
        3.2.4 Change Control Boards (CCBs)
        3.2.5 Interfaces
        3.2.6 Level of control
        3.2.7 Document revisions
        3.2.8 Automated tools used to perform change control
    3.3 Configuration Status Accounting
        3.3.1 Storage, handling and release of project media
        3.3.2 Information and Control
        3.3.3 Reporting
        3.3.4 **Release process**
        3.3.5 Document status accounting
        3.3.6 Change management status accounting
    3.4 Configuration Auditing
        3.4.1 Internal Audit
        3.4.2 Configuration Audit
        3.4.3 Other reviews
4.0 CM MILESTONES
5.0 TRAINING
6.0 SUBCONTRACTOR/VENDOR SUPPORT
```

Figure 5-5 Example SCM Plan.

CMMI®-SW (Staged) SP1.1. Identify configuration items. Identify the configuration items, components, and related work products that will be placed under configuration management.

IEEE Std 828 Section 4.3.1, Configuration Identification. Configuration identification activities shall identify, name, and describe the documented physical and functional characteristics of the code, specifications, design, and data elements to be controlled for the project. The documents are acquired for configuration control. Controlled items may be intermediate and final outputs (such as executable code, source code, user documentation, program listings, databases, test cases, test plans, specifications, and management plans) and elements of the support environment (such as compilers, operating systems, programming tools, and test beds).

The plan identifies the project configuration items (CI) and their structures at each project control point. It states how each CI and its versions are to be uniquely named and describes the activities performed to define, track, store, and retrieve CIs.

The Goals of Software Configuration Management Revisited

Configuration management refers to product control. This applies to the final deliverable and to all interim project artifacts. Each software product has a number of associated components; these components can have many versions. All versions of these components and the resulting product must be controlled.

SW-CMM® Goals for Software Configuration Management

Goal 1. SCM activities are planned. As described in the SCM Plan, IEEE Std 828, this standard addresses all levels of expertise, the entire life cycle, other organizations, and the relationships to hardware and other activities on a project. It is also not restricted to any form, type, or class of software. Other strengths of this standard are the special attention it pays to interface control and subcontractor/vendor control, and the extensive lists of items it provides for consideration in each key component area. Regarding interface control, this standard provides a list of possible interfaces and a minimum amount of information that must be defined for each interface. Regarding subcontractor/vendor control, this standard provides a list of information that must be addressed for subcontracted and acquired software.

Goal 2. Selected software work products are identified, controlled, and available. IEEE Std 828 supports the definition, documentation, and control of software project work products. IEEE 828 describes how to document the code baseline, all associated work products, and project documentation.

Goal 3. Changes to identified software work products are controlled. IEEE Std 828 then also describes how changes to these baselines are controlled.

Goal 4. Affected groups and individuals are informed of the status and content of software baselines. Status accounting and reporting are addressed in IEEE Std 828 in detail and supplemented by IEEE Std 730, Software Quality Assurance. The goal being not to show the value of SCM activities, but rather facilitate making controlled products available.

Although this standard provides excellent guidance on SCM plan creation and tailoring, there were a few weaknesses associated with it. The application of this standard would be much more effective for users if it provided a sample CM plan in conjunction with the plan development criteria. The standard could benefit by the inclusion a sample flowchart of a simple change control process addressing the life cycle at the component level; this would provide users with clarification on how certain CM activities may change throughout the life cycle.

CMMI®-SW (Staged) Goals for Configuration Management

The CMMI®-SW (Staged) emphasizes the need to place acquired products under configuration management by both the supplier and the project. If required, this can be supported by IEEE Std 1062-1998, Recommended Practice for Software Acquisition. Configuration management of work products may be performed at several levels of granularity. Provisions for conducting configuration management should be established in supplier agreements. As previously described in the SW-CMM® analysis of the SCM KPA, methods to ensure that the data is complete and consistent should be established and maintained.

- SG 1. Establish baselines. Baselines of identified work products are established.
- SG 2. Track and control changes. Changes to the work products under configuration management are tracked and controlled.
- SG 3. Establish integrity. Integrity of baselines is established and maintained.
- CG2. Institutionalize a managed process. The process is institutionalized as a managed process.

The CMMI®-SW (Staged) fully supports the requirement to establish and maintain a plan for performing all configuration management process activities. This emphasis is different in focus from the SW-CMM® requirement to have a documented SCM plan at the project level. The CMMI®-SW also requires the documentation of project-level SCM activities, but also requires the description of organizational SCM activities. IEEE Std 828-1998, IEEE Standard for Software Configuration Management Plans, can be used to help support this requirement.

SOFTWARE SUBCONTRACT/SUPPLIER MANAGEMENT

The Software Subcontractor Management KPA is about selecting and managing suppliers. Commitments between prime and subcontractor are negotiated and agreed to, those commitments are recorded, and the prime tracks subcontractor performance against these agreed commitments. This KPA can be considered to be the application to subcontractors of the other process areas of Level 2.

The Goals for CMM® Software Subcontractor Management

Software subcontract management can be thought of as the Level 2 management of subcontractors. The subcontract management KPA maps well to appropriate parts of ISO

9001; refer to section 4.6 on Purchasing (specifically section 4.6.2 Assessment of subcontractors).

Commitments between the prime and subcontractor are negotiated and agreed to and the commitments are recorded. During the lifetime of the contract, the prime and subcontractor keep in communication. The prime is responsible for tracking the performance and status of the subcontractor against the recorded commitments.

> Goal 1. The prime contractor selects qualified software subcontractors.
>
> Goal 2. The prime contractor and the software subcontractor agree to their commitments to each other.
>
> Goal 3. The prime contractor and the software subcontractor maintain ongoing communications.
>
> Goal 4. The prime contractor tracks the software subcontractor's actual results and performance against its commitments.

Software subcontract management involves selecting a software subcontractor, establishing commitments with the subcontractor, and tracking and reviewing the subcontractor's performance and results. These practices cover the management of a software subcontract, as well as the management of the software component of a subcontract that includes software, hardware, and possibly other system components. The subcontractor is selected based on its ability to perform the work.

Many factors contribute to the decision to subcontract a portion of the prime contractor's work. Subcontractors may be selected based on strategic business alliances, as well as technical considerations. The practices of this key process area address the traditional acquisition process associated with subcontracting a defined portion of the work to another organization.

When subcontracting, a documented agreement covering the technical and nontechnical (e.g., schedule) requirements is established and is used as the basis for managing the subcontract. The work to be done by the subcontractor and the plans for the work are documented. The standards that are to be followed by the subcontractor must be compatible with the prime contractor's standards.

The subcontractor performs all software planning, tracking, and oversight activities for the subcontracted work. The prime contractor ensures that these planning, tracking, and oversight activities are performed appropriately and that the software products delivered by the subcontractor satisfy their acceptance criteria. The prime contractor works with the subcontractor to manage their product and process interfaces.

The Goals for CMMI®-SW (Staged) Supplier Agreement Management

This CMMI-SW® KPA is concerned with product acquisition. The caveat here is that a formal agreement is required in support of these acquisition activities. In addition, the CMMI-SW® requires that supplier selection criteria for each acquisition type be identified.

> SG 1. Establish supplier agreements. Agreements with the suppliers are established and maintained.

SG2. Satisfy supplier agreements. Agreements with the suppliers are satisfied by both the project and the supplier.

CG2. Institutionalize a managed process.

Supplier agreement management requires effective coverage for all aspects of the software acquisition life cycle from supplier product review, agreement execution, product acceptance, and product transition. Supplier agreement management provides an extension of project planning providing for those activities that require reliance on third-party vendor support for successful project completion. This KPA does not directly address integrated teams, in which product suppliers are integrated members of the development team, but can provide support in defining those relationships if they exist.

Supporting IEEE Software Engineering Standards

IEEE Recommended Practice for Software Acquisition IEEE Std 1062-1998.

This standard provides information on the recommended practice for acquiring software. It describes a set of quality practices that can be applied during one or more steps of the software acquisition process.

This practice can be applied to software that runs on any platform, regardless of size, complexity, or criticality. It introduces the software acquisition life cycle, the nine steps in acquiring quality software, and steps for identifying potential suppliers. It offers support in preparing contract requirements, proposal evaluation, and supplier selection. It provides insight into the management of a software supplier and product acceptance.

This standard also offers a series of checklists (Annex A), which consist of information designed to help organizations establish their own software acquisition process (see Table 5-6).

The following matrix provides a cross-reference of the SEI CMM® Level 2 Software Subcontractor Management KPA to relevant supporting documentation from the IEEE Software Engineering Standards Collection. This is not meant to be an exhaustive list; rather, it is used to illustrate how the various KPAs are directly supported by IEEE Standards.

Table 5-6 IEEE Std 1062-1998 for software acquisition—checklists

Checklist 1: Organizational strategy
Checklist 2: Define the software
Checklist 3: Supplier evaluation
Checklist 4: Supplier and acquirer obligations
Checklist 5: Quality and maintenance plans
Checklist 6: User survey
Checklist 7: Supplier performance standards
Checklist 8: Contract payments
Checklist 9: Monitor supplier progress
Checklist 10: Software evaluation
Checklist 11: Software test
Checklist 12: Software acceptance

IEEE/CMM® Software Subcontractor Management Matrix

CMM® Goals and Key Practices	Definiton	IEEE Source	Comments
G1	The prime contractor selects qualified software subcontractors.		
AC1	The work to be subcontracted is defined and planned according to a documented procedure.	IEEE Std 1062-1998	Section 5.3.1, define the software being acquired.
AC2	The software subcontractor is selected, based on an evaluation of the subcontract bidders' ability to perform the work, according to a documented procedure.	IEEE Std 1062-1998	Section 5.3.2, establish proposal evaluation standards; Section 5.3.3, establish acquirer and supplier obligations; Section 5.3.4, develop plans to evaluate and accept software and services; Section 5.4, identifying potential suppliers.
G2	The prime contractor and the software subcontractor agree to their commitments to each other.		
AC3	The contractual agreement between the prime contractor and the software subcontractor is used as the basis for managing the subcontract.	IEEE Std 1062-1998	Section 5.5, preparing contract requirements; Section 5.7, managing for supplier performance.
AC4	A documented subcontractor's software development plan is reviewed and approved by the prime contractor.	IEEE Std 1062-1998	Section 5.6, evaluating proposal and selecting a supplier; Section 5.7.1, manage the contract during execution; Section 5.7.2, monitor supplier's progress.
	Note: The standard states that "some means" for performance feedback is required, but does not specifically require review of SDP.	IEEE Std 1058-1998	Section 4.7.7, subcontractor management plans (subclause 7.7 of SPMP).
AC5	A documented and approved subcontractor's software development plan is used for tracking the software activities and communicating status.	IEEE Std 1058-1998	Section 4.7.7, subcontractor management plans (subclause 7.7 of SPMP).
		IEEE Std 1062-1998	Section 5.6, evaluating proposal and selecting a supplier; Section 5.7.1, manage the contract during execution; Section 5.7.2, monitor supplier's progress.
	Note: There is a requirement to use a tailored version of the IEEE Std 1058-1998.		
AC6	Changes to the software subcontractor's statement of work, subcontract terms and conditions, and other commitments are resolved according to a documented procedure.	IEEE Std 1062-1998	Section 5.7.1, manage the contract during execution; Section 5.7.2, monitor supplier's progress.

IEEE/CMM® Software Subcontractor Management Matrix (*continued*)

CMM® Goals and Key Practices	Definiton	IEEE Source	Comments
G3	The prime contractor and the software subcontractor maintain ongoing communications.		
AC7	The prime contractor's management conducts periodic status/ coordination reviews with the software subcontractor's management.	IEEE Std 1062-1998 IEEE Std 1058-1998	Section 5.7.2, monitor supplier's progress, and annex A, checklist 9. Section 4.7.7, subcontractor management plans (subclause 7.7 of SPMP).
AC8	Periodic technical reviews and interchanges are held with the software subcontractor. Note: It is pointed out in the standard that a representative should be appointed to deal with subcontractors on contract issues, not technical issues/reviews.	IEEE Std 1062-1998 IEEE Std 1058-1998	If the technical feasibility of a "work segment" is deemed important, this could be a review criterion, but not a requirement. Section 4.7.7, subcontractor management plans (subclause 7.7 of SPMP).
AC9	Formal reviews to address the subcontractor's software engineering accomplishments and results are conducted at selected milestones according to a documented procedure.	IEEE Std 1062-1998 IEEE Std 1058-1998	Section 5.7.2, monitor supplier's progress. Section 4.7.7, subcontractor management plans (subclause 7.7 of SPMP).
AC13	The software subcontractor's performance is evaluated on a periodic basis, and the evaluation is reviewed with the subcontractor.	IEEE Std 1062-1998 IEEE Std 1058-1998	Section 5.7.2, monitor supplier's progress. Section 4.7.7, subcontractor management plans (subclause 7.7 of SPMP).
G4	The prime contractor tracks the software subcontractor's actual results and performance against its commitments.		
AC3	The contractual agreement between the prime contractor and the software subcontractor is used as the basis for managing the subcontract.	IEEE Std 1062-1998	Section 5.7.1, manage the contract during execution.
AC5	A documented and approved subcontractor's software development plan is used for tracking the software activities and communicating status.	IEEE Std 1062-1998 IEEE Std 1058-1998	Section 5.7.1, manage the contract during execution. Section 4.7.7, subcontractor management plans (subclause 7.7 of SPMP).

IEEE/CMM® Software Subcontractor Management Matrix (*continued*)

CMM® Goals and Key Practices	Definiton	IEEE Source	Comments
AC7	The prime contractor's management conducts periodic status/coordination with the software subcontractor's management.	IEEE Std 1062-1998	Section 5.7, managing for supplier performance.
AC9	Formal reviews to address the subcontractor's software engineering accomplishments and results are conducted at selected milestones according to a documented procedure.	IEEE Std 1062-1998	Section 5.7, calls for reviews at milestones, review of work product and acceptance, but no specification of the type of review.
AC10	The prime contractor's software quality assurance group monitors the subcontractor's software quality assurance activities according to a documented procedure—"software quality assurance group; monitors subcontractor's quality assurance; monitors the subcontractor's software quality assurance activities according to a documented procedure; monitors subcontractor's software quality assurance."	IEEE Std 730-2002	Section 4.12, supplier control.
AC11	The prime contractor's software configuration management group monitors the subcontractor's activities for software configuration management according to a documented procedure—"software configuration management group: monitors subcontractor's software configuration management; monitors the subcontractor's activities for software configuration management according to a documented procedure."	IEEE Std 828-1998	Section 4.3.6, subcontractor/vendor control.

IEEE/CMM® Software Subcontractor Management Matrix (*continued*)

CMM® Goals and Key Practices	Definiton	IEEE Source	Comments
AC12	The prime contractor conducts acceptance testing as part of the delivery of the subcontractor's software products according to a documented procedure—"acceptance testing: subcontractor's products; as part of the delivery of the subcontractor's software products according to a documented procedure: acceptance testing of subcontractor's products."	IEEE Std 1062-1998	Section 5.8, accepting the software; Section 5.8.1, evaluate and test the software; Section 5.8.2, maintain control over the test; Section 5.8.3, establish an acceptance process.
AC13	The software subcontractor's performance is evaluated on a periodic basis, and the evaluation is reviewed with the subcontractor.	IEEE Std 1062-1998	Section 5.7.2, monitor supplier's progress.
CO1	The project follows a written organizational policy for managing the software subcontract.	IEEE Std 1062-1998	Section 5.1, planning organizational strategy.
CO2	A subcontract manager is designated to be responsible for establishing and managing the software subcontract.	IEEE Std 1062-1998	Section 5.2, implementing organization's process; Section 5.2.1, establish a software acquisition process.
AB1	Adequate resources and funding are provided for selecting the software subcontractor and managing the subcontract.	IEEE Std 1062-1998	Section 5.2, implementing organization's process.
AB2	Software managers and other individuals who are involved in establishing and managing the software subcontract are trained to perform these activities.	IEEE Std 1062-1998 IEEE Std 1058-1998	Section 5.2.2, Item F. Section 4.5.1.4, project staff training plan.
AB3	Software managers and other individuals who are involved in managing the software subcontract receive orientation in the technical aspects of the subcontract.	IEEE Std 1062-1998 IEEE Std 1058-1998	Section 5.2.2, Item F. Section 4.5.1.4, project staff training plan.

IEEE/CMM® Software Subcontractor Management Matrix (*continued*)

CMM® Goals and Key Practices	Definiton	IEEE Source	Comments
ME1	Measurements are made and used to determine the status of the activities for managing the software subcontract. Note: This is commonly found to be a weak area during assessments; project tracking and oversight should be addressed more vigilantly, specifically in IEEE Std 1058-1998.	IEEE Std 1058-1998 IEEE Std 1061-1998	Section 4.5.3.6, metrics collection plan. Software quality metrics methodology.
VE1	The activities for managing the software subcontract are reviewed with senior management on a periodic basis.	IEEE Std 1058-1998	Section 4.5.3.5, reporting plan; Section 4.7.5, reviews and audits plan.
VE2	The activities for managing the software subcontract are reviewed with the project manager on both a periodic and event-driven basis.	IEEE Std 1058-1998	Section 4.5.3.5, reporting plan; Section 4.7.5, reviews and audits plan.
VE3	The software quality assurance group reviews and/or audits the activities and work products for managing the software subcontract and reports the results.	IEEE Std 730-2002	Section 4.12, supplier control.

IEEE/CMMI®-SW (Staged) Supplier Agreement Management Matrix

CMMI® Goals and Processes	Definition	IEEE Source	Comments
SG1	Establish supplier agreements.		
SP 1.1	Determine acquisition type. Determine the type of acquisition for each product or product component to be acquired.	IEEE Std 1062-1998	Section 5.3.1, define the software being acquired.
SP 1.2	Select suppliers. Select suppliers based on an evaluation of their ability to meet the specified requirements and established criteria.	IEEE Std 1062-1998	Section 5.3.2, establish proposal evaluation standards; Section 5.3.3, establish acquirer and supplier obligations; Section 5.3.4, develop plans to evaluate and accept software and services; Section 5.4, identifying potential suppliers.

IEEE/CMMI®-SW (Staged) Supplier Agreement Management Matrix (continued)

CMMI® Goals and Processes	Definition	IEEE Source	Comments
SP 1.3	Establish supplier agreements. Establish and maintain formal agreements with the supplier.	IEEE Std 1062-1998	Section 5.5, preparing contract requirements; Section 5.7, managing for supplier performance.
SG2	Satisfy supplier agreements.		
SP 2.1	Review COTS products. Review candidate COTS products to ensure they satisfy the specified requirements that are covered under a supplier agreement.	IEEE Std 1062-1998	Section 5.3.1, define the software being acquired; Section 5.3.2, establish proposal evaluation standards; Section 5.3.3, establish acquirer and supplier obligations; Section 5.3.4, develop plans to evaluate and accept software and services.
SP 2.2	Execute the supplier agreement. Perform activities with the supplier as specified in the supplier agreement.	IEEE Std 1062-1998	Section 5.7.1, manage the contract during execution; Section 5.7.2, monitor supplier's progress.
SP 2.3	Accept the acquired product. Ensure that the supplier agreement is satisfied before accepting the acquired product.	IEEE Std 1062-1998	Section 5.8, accepting the software; Section 5.8.1, evaluate and test the software; Section 5.8.2, maintain control over the test; Section 5.8.3, establish an acceptance process.
SP 2.4	Transition products. Transition the acquired products from the supplier to the project.	IEEE Std 1062-1998	Section 5.8, accepting the software; Section 5.8.3, establish an acceptance process.
GG2	Institutionalize a managed process.		
GP 2.1 (CO1)	Establish an organizational policy. Establish and maintain an organizational policy for planning and performing the supplier agreement management process.	IEEE Std 1062-1998	Section 5.1, planning organizational strategy.
GP 2.2 (AB1)	Plan the process. Establish and maintain the plan for performing the supplier agreement management process.	IEEE Std 1062-1998	Section 5.6, evaluating proposal and selecting a supplier; Section 5.7.1, manage the contract during execution; Section 5.7.2, monitor supplier's progress.
		IEEE Std 1058-1998	Section 4.7.7, subcontractor management plans (subclause 7.7 of SPMP).
GP 2.3 (AB 2)	Provide resources. Provide adequate resources for performing the supplier agreement management process, developing the work products, and providing the of the process.	IEEE Std 1062-1998	Section 5.2, implementing organization's process.

IEEE/CMMI®-SW (Staged) Supplier Agreement Management Matrix (continued)

CMMI® Goals and Processes	Definition	IEEE Source	Comments
GP 2.4 (AB3)	Assign responsibility. Assign responsibility and authority for performing the process, developing the work products, and providing the services of the supplier agreement management process.	IEEE Std 1062-1998	Section 5.2, implementing organization's process; Section 5.2.1, establish a software acquisition process.
GP 2.5 (AB4)	Train people. Train the people performing or supporting the supplier agreement management process as needed.	IEEE Std 1058-1998	Section 4.5.1.4, project staff training plan; section calls for a training plan, up to the PM to include this activity training.
GP 2.6 (DI1)	Manage configurations. Place designated work products of the supplier agreement management process under appropriate levels of configuration management.	IEEE Std 828-1998	Section 4.3.6, subcontractor/vendor control.
GP 2.7 (DI2)	Identify and involve relevant stakeholders. Identify and involve the relevant stakeholders of the supplier agreement management process as planned.	IEEE Std 1058-1998	Section 4.5.3.5, reporting plan; Section 4.7.5, reviews and audits plan.
GP 2.8 (DI3)	Monitor and control the process. Monitor and control the supplier agreement management process against the plan for performing the process and take appropriate corrective action.	IEEE Std 1058-1998	Section 4.5.3.5, reporting plan; Section 4.7.5, reviews and audits plan; Section 5.5, preparing contract requirements; Section 5.7, managing for supplier performance.
GP 2.9 (VE 1)	Objectively evaluate adherence. Objectively evaluate adherence of the supplier agreement management process against its process description, standards, and procedures, and address noncompliance.	IEEE Std 730-2002	Section 4.12, supplier control.
GP 2.10 (VE2)	Review status with higher-level management. Review the activities, status, and results of the supplier agreement management process with higher-level management and resolve issues.	IEEE Std 1058-1998	Section 4.5.3.5, reporting plan; Section 4.7.5, reviews and audits plan.

SW-CMM® Software Subcontractor Management Analysis

There is no obvious IEEE requirement to have a software development plan for each subcontractor; this is the CMM® Level 2 basis for monitoring the progress of the supplier. If you dig, you find the requirement in Section 4.7.7, *Subcontractor management plans,* of the 1058 Standard. This section states:

> The SPMP shall contain plans for selecting and managing any subcontractors that may contribute work products to the software project. **The criteria for selecting subcontractors shall be specified and the management plan for each subcontract shall be generated using a tailored version of this standard.** Tailored plans should include the monitoring of technical progress, schedule and budget control, product acceptance criteria, and risk management procedures shall be included in each subcontractor plan. Additional topics should be added as needed to ensure successful completion of the subcontract. A reference to the official subcontract and prime contractor/subcontractor points of contact shall be specified.

IEEE Std 1058-1998, *IEEE Standard for Software Project Management Plans,* Section 4.7.7, *Subcontractor management plans,* and IEEE Std 1062-1998, *IEEE Recommended Practice for Software Acquisition* each provide valuable insight into the items required for effective subcontractor elicitation, evaluation, acquisition, and management. This section states generally what is required and states that a tailored version of Std 1058 should be used for subcontracted software management plans. However, more information would benefit the standards users.

CMMI®-SW (Staged) Supplier Agreement Management Analysis

The CMMI®-SW requires the definition of the acquisition process and the requirements for each product to be acquired. It also incorporates the requirement to review candidate COTS products to ensure they satisfy the specified requirements covered in the supplier agreement. IEEE Std 1062 can be used as a resource in support of these requirements. This standard describes the software acquisition life cycle and supporting processes in detail (see Table 5-7). This standard includes a checklist that may be used by organizations

Table 5-7 IEEE Std 1062 description of software acquisition milestones [25]

Phase	Phase initiation milestone	Phase completion milestone	Steps in software acquisition process
Planning	Develop idea	Release the RFP	Planning organizational strategy, implementing organization's process, and determining software requirements
Contracting	RFP is released	Contract is signed	Identifying potential suppliers, preparing contract requirements, and evaluating proposals and selecting the supplier
Product implementation	Sign contract	The software product is received	Managing supplier performance
Follow-on	Accept software product	Product is no longer in use	Using the software

when establishing a software acquisition process and detailed guidelines for supporting acquisition planning.

The CMMI®-SW is more rigorous in its requirements in support of acquisition plan maintenance. IEEE Std 1062 provides guidance in support of the documentation required to support the SW-CMM® KPA for Software Subcontractor Management and the CMMI®-SW (Staged) KPA for Supplier Agreement Management (see Table 5-8). Appendix B of this IEEE Software Engineering standard provides a suggested document format with detailed guideline support for each recommended document section.

Example of IEEE KPA Support for Software Subcontractor Management

CMM® AC1. The work to be subcontracted is defined and planned according to a documented procedure.

CMMI®-SW (Staged) SP 1.1. Determine acquisition type. Determine the type of acquisition for each product or product component to be acquired.

IEEE Std 1062 Section 5.3 Defining the Software Requirements. 5.3.1 Define the software being acquired. The objective is to obtain from the supplier(s) realistic assessments of the size, scope, and cost of the effort required to produce the software.

Table 5-8 Suggested acquisition plan documentation format [17]

1. Introduction
2. References
3. Definitions
4. Software acquisition overview
    4.1 Organization
    4.2 Schedule
    4.3 Resource summary
    4.4 Responsibilities
    4.5 Tools, techniques, and methods
5. Software acquisition process
    5.1 Planning organizational strategy
    5.2 Implementing the organization's process
    5.3 Determining the software requirements
    5.4 Identifying potential suppliers
    5.5 Preparing contract documents
    5.6 Evaluating proposals and selecting the suppliers
    5.7 Managing supplier performance
    5.8 Accepting the software
    5.9 Using the software
6. Software acquisition reporting requirements
7. Software acquisition management requirements
    7.1 Anomaly resolution and reporting
    7.2 Deviation policy
    7.3 Control procedures
    7.4 Standards, practices, and conventions
    7.5 Performance tracking
    7.6 Quality control of plan
8. Software acquisition documentation requirements

The needed software, deliverables, and software support should be described as completely as possible in the RFP so that the supplier can understand and address the scope of work in the proposal. The example questions in Annex A, checklist 2, may be used as a starting point. For fully developed software, IEEE Std 830-1998 should be used to document the requirements. Depending upon the type of software being acquired, a request for quote or other acquisition document may be used in place of the RFP.

The Goals of Software Subcontractor Management Revisited

SW-CMM® Goals for Software Subcontractor Management

Goal 1. The prime contractor selects qualified software subcontractors. The software estimates of Goal 1 are for size, schedule, effort, and so on. These estimates are made, documented, and used in the software development plan to communicate commitment.

Goal 2. The prime contractor and the software subcontractor agree to their commitments to each other.

Goal 3. The prime contractor and the software subcontractor maintain ongoing communications.

Goal 4. The prime contractor tracks the software subcontractor's actual results and performance against its commitments.

IEEE Std 1058, the standard in support of software project management planning, can be used to support this KPA, just at it is used in support of the Software Project Planning KPA. An additional standard, IEEE Std 1063, Recommended Practice for Software Acquisition, should also be used in support of meeting these requirements. It is important to note that an older standard, EIA/IEEE J-STD-016-1995 Standard for Information Technology Software Life Cycle Processes Software Development—Acquirer-Supplier Agreement, is also a good source of material as well. This standard is not included in the current standards set, or in the previous matrix mapping, but is still available from IEEE Press.

CMMI®-SW (Staged) Goals for Supplier Agreement Management. IEEE Std 1063 supports the analysis and definition of the processes supporting the software acquisition life cycle. Complete guidance for the CMMI®-SW KPA Supplier Agreement Management is provided when used in conjunction with the IEEE standard supporting software project planning, IEEE Std 1058.

SG1. Establish supplier agreements. Agreements with the suppliers are established and maintained.

SG2. Satisfy supplier agreements. Agreements with the suppliers are satisfied by both the project and the supplier.

CG2. Institutionalize a managed process. The process is institutionalized as a managed process.

A clear description of how adequately IEEE Std 1063 supports the CMMI®-SW for Supplier Agreement Management can be seen in a description of the nine steps for software acquisition that is provided within the standard (see Table 5-9). IEEE Std 1063 supports each step with prescriptive detail.

Table 5-9 IEEE Std 1063 Steps for software acquisition [25]

Step 1	Planning organizational strategy. Review acquirer's objectives and develop a strategy for acquiring software.
Step 2	Implementing organization's process. Establish a software acquisition process that fits organization's needs for obtaining a quality software product. Include appropriate contracting practices.
Step 3	Determining the software requirements. Define the software being acquired and prepare quality and maintenance plans for accepting software supplied by the supplier.
Step 4	Identifying potential suppliers. Select potential candidates who will provide documentation for their software, demonstrate their software, and provide formal proposals. Failure to perform any of these actions is basis to reject a potential supplier. Review supplier performance data from previous contracts.
Step 5	Preparing contract requirements. Describe the quality of the work to be done in terms of acceptable performance and acceptance criteria, and prepare contract provisions that tie payments to deliverables. Review contract with legal counsel.
Step 6	Evaluating proposals and selecting the supplier. Evaluate supplier proposals, select a qualified supplier, and negotiate the contract. Negotiate with an alternate supplier, if necessary.
Step 7	Managing supplier performance. Monitor supplier's progress to ensure all milestones are met and to approve work segments. Provide all acquirer deliverables to the supplier when required.
Step 8	Accepting the software. Perform adequate testing and establish a process for certifying that all discrepancies have been corrected and that all acceptance criteria have been satisfied.
Step 9	Using the software. Conduct a follow-up analysis of the software acquisition contract to evaluate contracting practices, record lessons learned, and evaluate user satisfaction with the product. Retain supplier performance data.

MEASUREMENT AND ANALYSIS

The Goals of CMMI® Measurement and Analysis

Software measurement and analysis activities help to determine whether to whether the software quality requirements are being met. There are two goals when discussing measurement and analysis: the process goal and the product goal. The process goal is to provide measures that may be applicable throughout the life cycle and may provide the means for continual self-assessment and reliability improvement. The product goal is to increase the overall reliability of the developed software.

> SG1. Align measurement and analysis activities. Measurement objectives and activities are aligned with identified information needs and objectives.
>
> SG2. Provide measurement results. Measurement results that address identified information needs and objectives are provided.
>
> CG2. Institutionalize a managed process. The process is institutionalized as a managed process.

Supporting IEEE Software Engineering Standards

IEEE Standard for Developing Software Life Cycle Processes IEEE Std 1074™-1997. The IEEE Standard for Developing Software Life Cycle Processes (SLCP), IEEE Std 1074, provides for the definition and maintenance of the processes that govern a software project. This standard applies to the management and support activities throughout the entire software lifecycle. It also provides guidance in all aspects of the software life cycle from concept exploration through retirement.

IEEE Std 1074 can help process architects develop the SLCP that is required in support of a specific software project. This standard does not support the development of organizational software life cycle processes. If support of organizational process development is needed, organizations should take advantage of the information provided in the IEEE 12207 series: Software Life Cycle Processes (12207.0), Life Cycle Data (12207.1), and Implementation Considerations (12207.2).

Annex A of Standard 1074 describes all the required activities in support of SLCP development, including the definition and evaluation of project metrics. Annex B of this standard provides an example of SCLP development. Annex C provides an informative mapping template that can be used as a checklist to identify and track required project deliverable.

IEEE Standard IEEE Standard Dictionary of Measures to Produce Reliable Software IEEE Std 982.1. This standard provides a set of software reliability measures that can be applied to the software product as well as to the supporting development processes. This standard provides measures that can be applied early in the development process (see Table 5-10). It provides a common set of measures and is designed to assist management in product management and oversight activities. The standard describes and supports measure for both software product and process.

IEEE Standard Classification for Software Anomalies IEEE Std 1044™-1993 (R2002). IEEE Std 1044, Standard Classification for Software Anomalies, defines a uniform approach to the classification and documentation of the variances found in software products. The goal in applying this standard is that as software problems are discovered and reported, through the use of this classification scheme, issues with both the product and the supporting software life cycle can be discovered and improved.

This classification methodology can be used to classify a problem with either product or process. This standard provides support from the initial recognition of the anomaly to its final disposition. It supports the customization of the classification process due to specific organizational requirements.

IEEE Standard for Software Productivity Metrics IEEE Std 1045-1992 (R2003). This standard provides a framework for measuring and reporting software productivity. It focuses on how to measure software productivity and reporting. It is meant for those who want to measure the productivity of the software process in support of their software product. Effective productivity measures can provide valuable insight into the associated software processes. Interpreting productivity based on a single number leaves much unknown about the process being measured. IEEE Std 1045 requires

Table 5-10 IEEE Std 982.1, List of Measures for Reliable Software

Paragraph	Description of Measure
4.1	Fault Density
4.2	Defect Density
4.3	Cumulative Failure Profile
4.4	Fault-Days Number
4.5	Functional or Modular Test Coverage
4.6	Cause and Effect Graphics
4.7	Requirements Traceability
4.8	Defect Indices
4.9	Error Distribution
4.10	Software Maturity Index
4.11	Man hours per Major Defect Detected
4.12	Number of Conflicting Requirements
4.13	Number of Entries and Exits per Module
4.14	Software Science Measures
4.15	Graph-Theoretic Complexity for Architecture
4.16	Cyclomatic Complexity
4.17	Minimal Unit Test Case Determination
4.18	Run Reliability
4.19	Design Structure
4.20	Mean Time to Discover the Next K Faults
4.21	Software purity Level
4.22	Estimated Number of Faults Remaining
4.23	Requirement Compliance
4.24	Test Coverage
4.25	Data or Information Flow Complexity
4.26	Reliability Growth Function
4.27	Residual Fault Count
4.28	Failure Analysis Using Elapsed Time
4.29	Testing Sufficiency
4.30	Mean-Time-to-Failure
4.31	Failure Rate
4.32	Software Documentation and Source Listings
4.33	Required Software Reliability
4.34	Software Release Readiness
4.35	Completeness
4.36	Test Accuracy
4.37	System Performance Reliability
4.38	Independent Process Reliability
4.39	Combined Hardware and Software Operational Availability

additional information, such as the scope, characteristics of the process being measured, and the precision of the data being used in the calculations. In those situations in which precise measurement definitions are not possible, this standard requests that descriptions of the supporting processes used and the measurements taken be done in a specified format.

IEEE/CMMI®-SW (Staged) Measurement and Analysis Matrix

CMMI® Goals and Processes	Definition	IEEE Source	Comments
SG1	Align measurement and analysis activities.		
SP 1.1	Establish measurement objectives. Establish and maintain measurement objectives that are derived from identified information needs and objectives.	IEEE Std 1061-1998	Section 4.2.2.2, identify the benefits of applying the metrics.
		IEEE Std 1074-1997	This is not specifically called out in the standard and should be added to Section A.1.1.4 as part of Input—Define objectives and as part of Output—Review objectives.
SP 1.2	Specify measures. Specify measures to address the measurement objectives.	IEEE Std 1074-1997	Section A.1.1.4, define metrics for life cycle.
		IEEE Std 1061-1998	Section 4.2, identify software quality metrics.
SP 1.3	Specify data collection and storage procedures. Specify how measurement data will be obtained and stored.	IEEE Std 1061-1998	Section 4.3.1, define the data collection procedures.
		IEEE Std 982.1-1988	Section 4, measures for reliable software.
SP 1.4	Specify analysis procedures. Specify how measurement data will be analyzed and reported.	IEEE Std 1044-1993	Section 4.1, classification process.
		IEEE Std 1061-1998	Section 4.3, implement the software quality metrics.
SG2	Provide measurement results.		
SP 2.1	Collect measurement data. Obtain specified measurement data.	IEEE Std 982.1-1988	Section 4, measures for reliable software.
		IEEE Std 1061-1998	Section 4.3.3, collect the data and compute the metric values; two tables are provided to define requirements for metrics set and data items.
SP 2.2	Analyze measurement data. Analyze and interpret measurement data.	IEEE Std 982.1-1988	Section 4, measures for reliable software.
		IEEE Std 1044-1993	Section 4.1, classification process.
		IEEE Std 1061-1998	Section 4.4, analyze the software metrics results.
SP 2.3	Store data and eesults. Manage and store measurement data, measurement specifications, and analysis results.	IEEE Std 1044-1993	Section 4.1, classification process. This standard addresses the storage of the results and supporting data, but not its management.

IEEE/CMMI®-SW (Staged) Measurement and Analysis Matrix (*continued*)

CMMI® Goals and Processes	Definition	IEEE Source	Comments
SP 2.4	Communicate results. Report results of measurement and analysis activities to all relevant stakeholders.	IEEE Std 982.1-1988	Section 4, measures for reliable software.
		IEEE Std 12207.1-1996	Section 6, specific information item content guidelines.
GG2	Institutionalize a managed process.		
GP 2.1 (CO1)	Establish an organizational policy. Establish and maintain an organizational policy for planning and performing the measurement and analysis process.	IEEE Std 1045-1992	Not specifically called out in the standard, but Annex A, Sample metrics data collection summary list, could be used as a basis for organizational policy.
GP 2.2 (AB1)	Plan the process. Establish and maintain the plan for performing the measurement and analysis process.	IEEE Std 1045-1992	Section 5, output primitives; Annex A, sample metrics data collection summary list.
		12207.1-1996	Section 6, specific information item content guidelines.
GP 2.3 (AB 2)	Provide resources. Provide adequate resources for performing the measurement and analysis process, developing the work products, and providing the services of the process.	IEEE Std 1061-1998	Section 4.2.2.1, identify the costs of implementing the metrics.
GP 2.4 (AB3)	Assign responsibility. Assign responsibility and authority for performing the process, developing the work products, and providing the services of the measurement and analysis process.	IEEE Std 1061-1998	Standard does not directly address issues as described, but in Section 4.2.3 requires the commitment of all parties involved. This is not the same as the documentation of responsibility, and should be added.
GP 2.5 (AB4)	Train people. Train the people performing or supporting the measurement and analysis process as needed.	IEEE Std 1061-1998	Standard does not directly address issues as described, and should be added.
		IEEE Std 1058-1998	Section 4.5.1.4, project staff training plan; include training activities in support of measurement and analysis activities.
GP 2.6 (DI1)	Manage configurations. Place designated work products of the measurement and analysis process under appropriate levels of configuration management.	IEEE Std 1061-1998	Standard does not directly address issues as described. The standard does require the documentation of the storage location.
		IEEE Std 828-1998	Use as basis for SCM planning.

IEEE/CMMI®-SW (Staged) Measurement and Analysis Matrix (*continued*)

CMMI® Goals and Processes	Definition	IEEE Source	Comments
GP 2.7 (DI2)	Identify and involve relevant stakeholders. Identify and involve the relevant stakeholders of the measurement and analysis process as planned.	IEEE Std 1061-1998	Standard does not directly address issues as described, but in Section 4.2.3 requires the commitment of all parties involved. This is not the same as the documentation of relevant stakeholders, and should be added.
GP 2.8 (DI3)	Monitor and control the process. Monitor and control the measurement and analysis process against the plan for performing the process and take appropriate corrective action.	IEEE Std 1058-1998	Section 4.5.3.6, metrics collection plan.
GP 2.9 (VE 1)	Objectively evaluate adherence. Objectively evaluate adherence of the measurement and analysis process against its process description, standards, and procedures, and address non-compliance.	IEEE Std 1058-1998 IEEE Std 730-2002	Section 4.7.4, quality assurance plan. Use as support document in SQA planning.
GP 2.10 (VE2)	Review status with higher-level management. Review the activities, status, and results of the measurement and analysis process with higher-level management and resolve issues.	IEEE Std 1058-1998	Section 4.5.3, control plan.

Analysis of Measurement and Analysis

IEEE Std 1058, Standard for Software Project Management Plans, points to the requirement of a metrics plan. This metrics plan must identify the specific measurement and analysis activity to be provided in support of a software effort. As previously described, there are a number of IEEE Standards supporting this KPA. In particular, the IEEE standard Classification for Software Anomalies (1044) and Standard for Productivity Metrics (1045) provide significant support for this KPA.

However, there are areas of weak support. The requirement to describe how the measurement and analysis data are stored and maintained should be addressed either in the metrics plan, or as a cross-reference to the project SCM plan. The identification of relevant stakeholders should also either be addressed in the metrics plan, or as a cross-reference to the project management plan. Responsibility and training are areas that should also be directly addressed during the documentation of the processes in support of this KPA.

Proposed Document Outline

Figure 5-6 provides a suggested format for a project level metrics plan. The metrics plan should contain a description of all measurement and analysis used in support of an identified software effort.

Example of IEEE KPA Support for Measurement and Analysis

> CMMI®-SW (Staged) SP 1.2. Specify Measures; Specify measures to address the measurement objectives.
> IEEE Std 1062, Section 4.2, Identify Software Quality Metrics.
> IEEE Std 1062, Section 4.3.1, Define the data collection procedures.

For each metric in the metrics set, determine the data that will be collected and determine the assumptions that will be made about the data (e.g., random sample and subjective or objective measure). Show the flow of data from point of collection to evaluation of metrics. Identify tools and describe how they will be used. Describe data storage procedures. Establish a traceability matrix between metrics and data items. Identify the organizational entities that will participate in data collection, including those responsible for monitoring data collection. Describe the training and experience required for data collection and the training process for personnel involved.

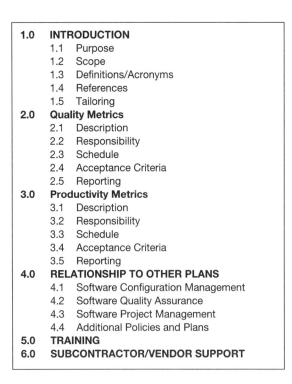

Figure 5-6 Example outline for metrics plan

The Goals of CMMI® Measurement and Analysis Revisited

Several IEEE software engineering standards support the definition and implementation of software metrics. These metrics support a uniform approach to the definition and measurement of productivity.

- SG1. Align measurement and analysis activities. Measurement objectives and activities are aligned with identified information needs and objectives.
- SG2. Provide measurement results. Measurement results that address identified information needs and objectives are provided.
- CG2. Institutionalize a managed process. The process is institutionalized as a managed process.

These goals are primarily supported by IEEE Std 1044, Standard Classification for Software Anomalies; IEEE Std 982.1, Standard Dictionary of Measures to Produce Reliable Software; IEEE Std 1045, Standard for Software Productivity Metrics; and IEEE Std 1061, Software Quality Metrics Methodology. These standards provide a common language and framework in support of software measurement and analysis activities. The information provided by these plans support the goals of the CMMI® Measurement and Analysis KPA as long as they are used in conjunction with the other IEEE standards supporting project planning, software configuration management, and software quality assurance.

6

Using IEEE Standards to Achieve Software Process Improvement

IEEE-SUPPORTED PROCESS IMPROVEMENT

From the analysis presented within this book, it should be readily apparent that IEEE software engineering standards (SES) can effectively be used to support the initiation of software process definition and improvement. It is time to revisit the IDEAL model (Figure 4-1, repeated on the next page for clarity) to describe the pairing of IEEE standards with CMM-based software process improvement.

Define and Train the Process Team (Initiate)

A process improvement team is a chosen group of people who are given responsibility and authority for improving a selected process in an organization; this team must have the backing of senior management. Process owners are responsible for the process design, not for the performance, of their associated process areas. The process owner is further responsible for the process measurement and feedback systems, the process documentation, and the training of the process performers in its structure and conduct. In essence, the process owner is the person ultimately responsible for improving a process.

Implementing process improvement can be very time-consuming, depending upon the scope and complexity of the process. Expectations for the process owner's time commitments and job responsibilities must be modified, accordingly, to reflect the new responsibilities. This commitment should reflect time budgeted for process definition and improvement and any required refresher training.

Software Engineering

Practically defined, software engineering is a discipline whose goal is the production of fault-free software, delivered on time and within budget, meeting customer expectations.

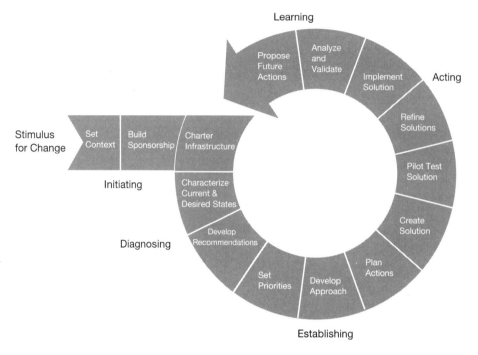

Figure 4-1 The IDEAL model [62].

IEEE Std 610.12, the Standard Glossary of Software Engineering Terminology, defines software engineering as:

(1) The application of a systematic, disciplined, quantifiable approach to the development, operation, and maintenance of software; that is, the application of engineering to software.
(2) The study of approaches as in (1). [2]

Traditionally, individuals graduating with computer science degrees have not been trained in a manner meeting this definition. In fact, except for universities that offer the senior-level undergraduate software engineering courses or masters' degrees in software engineering, computer science students have been typically trained in the various programming methodologies. Emphasis is too frequently placed upon the "craft" of programming rather than the processes that support software engineering.

Graduates are called computer scientists, but they are often performing engineering job functions without the benefit of formal engineering training. Many might argue that software engineering is a clearly defined profession, and recent graduates possess all the required skills in support of both programming and process. If this is a true statement, then why does the software process vary so drastically from organization to organization?

In order for software engineering to be truly defined as a profession, several questions need to be answered: What is software engineering's core body of knowledge? How should software engineers be trained? And how should software engineers be certified? These are the types of questions that can be answered by those practicing in other professional fields.

The IEEE Computer Society, through its Professional Practices Committee (PPC), has been working to provide the answers to the above questions. The PPC has been working to define the Software Engineering Body of Knowledge (SWEBOK) and software engineering certification through the development of the Certified Software Development Professional (CSDP) examination.

SWEBOK

On May 21, 1993, the IEEE Computer Society Board of Governors approved a motion to "establish a steering committee for evaluating, planning, and coordinating actions related to establishing software engineering as a profession." [44] Many professions are based on a body of knowledge that supports such purposes as accreditation of academic programs, development of education and training programs, certification of specialists, or professional licensing. One of the goals of this motion was to establish a body of knowledge in support of software engineering. In 1996, the initial Strawman version of this body of knowledge was published as the Software Engineering Body of Knowledge (SWEBOK[1]). The SWEBOK is now in its trial version and is being adopted by the IEEE Computer Society for publication.

The SWEBOK provides valuable guidance regarding what has become the consensus body of software engineering knowledge. The goal of the SWEBOK is to define the core of what the software engineering discipline should contain. The acknowledgment of the SWEBOK by educational institutions would mean a shift from technology-specific programming courses, like C++ and Java, to the knowledge and practices required in support of software development and software project management. The SWEBOK is divided into ten areas that contain descriptions and references to topically supportive material that reflect the knowledge required in support of the software engineering discipline (see Table 6-1).

The SWEBOK will continue to evolve, as do all the bodies of knowledge associated with other professional disciplines. Areas will be redefined and refined over time to address all of the factors that affect the creation or maintenance of a software product. The SWEBOK offers a challenge to all of those involved with software—broaden the knowledge base and move from computer science to software engineering. The requirement is for software engineering practitioners to not only know how to write code, but to understand all aspects of the processes that support the creation of their products.

IEEE and Software Engineering Training
In order for software process improvement to be successful, everyone involved should know how to effectively perform his or her roles. Many times, key individuals have not been effectively trained to perform to support the software engineering process. If people in your organization cannot answer the questions in Table 6-2 right away or if nobody knows what these roles are or who performs them, then your staff has an urgent need for some software engineering education.

IEEE software engineering standards provide support to the knowledge areas defined by the SWEBOK (see Table 6-3). Each standard contains additional reference material that may be used in support of software engineering training activities. Many of the specific questions that practitioners have may be answered by turning to the IEEE Software Engineering Standards Collection.

---

[1]SWEBOK is an official service mark of the IEEE.

Table 6-1 10 SWEBOK areas [44]

SWEBOK Area	Supports
Software Requirements	Requirements Engineering Process
	Requirements Elicitation
	Requirements Analysis
	Requirements Specification
	Requirements Validation
	Requirements Management
Software Design	Software Design Basic Concepts
	Key Issues in Software Design
	Software Structure and Architecture
	Software Design Quality Analysis and Evaluation
	Software Design Notation
	Software Design Strategies and Methods
Software Construction	Reduction in Complexity
	Anticipation of Diversity
	Structuring for Validation
	Use of External Standards
	Testing Basic Concepts and Definitions
	Test Levels
	Test Techniques
	Test-Related Measures
	Managing the Test Process
Software Maintenance	Basic Concepts
	Maintenance Process
	Key Issues in Software Maintenance
	Techniques for Maintenance
Software Configuration Management	Management of the SCM Process
	Software Configuration Identification
	Software Configuration Control
	Software Configuration Status Accounting
	Software Configuration Auditing
	Software Release Management and Delivery
Software Engineering Management	Organizational Management
	Process/Project Management
	Software Engineering Measurement
Software Engineering Process	Software Engineering Process Concepts
	Process Infrastructure
	Process Measurement
	Process Definition
	Qualitative Process Analysis
	Process Implementation and Change
Software Tools and Methods	Software Tools
	Software Requirements Tools
	Software Design Tools
	Software Construction Tools
	Software Testing Tools
	Software Maintenance Tools
	Software Engineering Process Tools
	Software Quality Tools
	Software Configuration Management

Table 6-1 *Continued*

SWEBOK Area	Supports

Software Tools and Methods (*cont.*)	Software Tools (*cont.*)
	    Tools
	    Software Engineering Management Tools
	    Infrastructure Support Tools
	    Miscellaneous Tool Issues
	Software Methods
	    Heuristic Methods
	    Formal Methods
	    Prototyping Methods
	    Miscellaneous Method Issues

Software Quality	Software Quality Concepts Definition and Planning for Quality Techniques Requiring Two or More People Support to Other Techniques Testing Special to SQA or V&V Defect Finding Techniques Measurement in Software Quality Analysis

Table 6-2 Key software engineering questions

To the project managers:
  a. What is the difference between a plan and a schedule?
  b. What do you record about the estimates that are being made?
  c. Do you estimate size as well as effort when doing your planning? Do you monitor both attributes during the life of the project?
To the configuration managers:
  a. What is a baseline?
  b. What is the purpose of a configuration audit?
  c. Who authorizes changes to the configuration units?
To the quality assurance analysts:
  a. What is the object of quality assurance?
  b. How is it different from quality control? From testing?
  c. Who, in the organization, knows about the quality assurance activities and results?

Certification

In 1995, the Software Engineering Institute (SEI) began to model the evolution and maturation of professions in order to understand how to facilitate the maturity of software engineering as a profession. In the resultant report, *A Mature Profession of Software Engineering* [51], it was observed that professionals follow professional-development paths that are fairly similar, regardless of their specific discipline, and that mature professions include the following elements:

[1] Initial professional education
[2] Accreditation
[3] Skills development

[4] Certification
[5] Licensing
[6] Professional development
[7] Professional societies
[8] Code of ethics

After completion of education and on-the-job skills development, many professionals are required to pass one or more exams. This is to certify that they are qualified to practice in their field, meeting some minimum level of competency. Individuals are also required to demonstrate continued growth and competency in their respective fields by continuing their education and recertification.

Based upon the criteria listed above, software engineering is growing into a profession throughout the world. In the United Kingdom, the Institution of Electrical Engineers conducts a software engineering certificate program and the British Computer Society has a professional development scheme. The Australian Computer Society offers a certification program in information technology with a subspecialty in software engineering.

Until recently, there has not been a certification program supporting the software engineering profession within the United States. The IEEE Computer Society began developing a certification program for software engineering practitioners in June 1999. This program has become known as the Certified Software Development Professional (CSDP) examiniation. The CSDP began testing applicants in 2002. The specifications for the CSDP are included in Table 6-4 as a software engineering training checklist.

Table 6-3 IEEE standards and training

Requirements	
How does one manage, elicit, and define requirements?	IEEE Std 830-1998, Recommended Practice for Software Requirements Specification
Test	
How does one classify software anomalies?	IEEE Std 1044-1993 (R2002), Standard Classification for Software Anomalies
How does one select and apply software measures?	IEEE Std 982.1-1988, Standard Dictionary of Measures to Produce Reliable Software
How does one define a test unit?	IEEE Std 1008-1987 (R2002), Standard for Software Unit Testing
What documentation is required in support of the testing process?	IEEE Std 829-1998, Standard for Software Test Documentation
Maintenance	
What maintenance activities are required prior to product delivery?	IEEE Std 1219, Standard for Software Maintenance
Communication	
How does one communicate using consistent terminology?	IEEE Std 610.12-1990 (R2002), Standard Glossary of Software Engineering Terminology
Configuration Management	
What describes the requirements and categories of information in support of configuration management planning?	IEEE Std 828-1998, Standard for Software Configuration Management Plans
Reviews and Audits	
Where does one find information describing software audit procedures?	IEEE Std 1028-1997 (2002), Standard for Software Reviews

Table 6-4 The IEEE Certified Software Development Professional (CSDP) Exam Specifications[1]

Business Practices and Engineering Economics (3–4% questions)
    A. Engineering Economics
    B. Ethics
    C. Professional Practice
    D. Standards

II. Software Requirements (13–15% questions)
    A. Requirements Engineering Process
    B. Requirements Elicitation
    C. Requirements Analysis
    D. Software Requirements Specification
    E. Requirements Validation
    F. Requirements Management

III. Software Design (22–24% questions)
    A. Software Design Concepts
    B. Software Architecture
    C. Software Design Quality Analysis and Evaluation
    D. Software Design Notations and Documentation
    E. Software Design Strategies and Methods
    F. Human Factors in Software Design
    G. Software and System Safety

IV. Software Construction (10–12% questions)
    A. Construction Planning
    B. Code Design
    C. Data Design and Management
    D. Error Processing
    E. Source Code Organization
    F. Code Documentation
    G. Construction QA
    H. System Integration and Deployment
    I. Code Tuning
    J. Construction Tools

V. Software Testing (15–17% questions)
    A. Types of Tests
    B. Test Levels
    C. Testing Strategies
    D. Test Design
    E. Test Coverage of Code
    F. Test Coverage of Specifications
    G. Test Execution
    H. Test Documentation
    I. Test Management

VI. Software Maintenance (3–5% questions)
    A. Software Maintainability
    B. Software Maintenance Process
    C. Software Maintenance Measurement
    D. Software Maintenance Planning
    E. Software Maintenance Management
    F. Software Maintenance Documentation

VII. Software Configuration Management (3–4% questions)
    A. Management of SCM Process
    B. Software Configuration Identification

(continued)

Table 6-4 *Continued*

VII. Software Configuration Management (3–4% questions) (*cont.*)
    C. Software Configuration Control
    D. Software Configuration Status Accounting
    E. Software Configuration Auditing
    F. Software Release Management and Delivery

VIII. Software Engineering Management (10–12% questions)
    A. Measurement
    B. Organizational Management and Coordination
    C. Initiation and Scope Definition
    D. Planning
    E. Software Acquisition
    F. Enactment
    G. Risk Management
    H. Review and Evaluation
    I. Project Close Out
    J. Post-closure Activities

IX. Software Engineering Process (2–4% questions)
    A. Process Infrastructure
    B. Process Measurement
    C. Process Definition
    D. Qualitative Process Analysis
    E. Process Implementation and Change

X. Software Engineering Tools and Methods (2–4% questions)
    A. Management Tools and Methods
    B. Development Tools and Methods
    C. Maintenance Tools and Methods
    D. Support Tools and Methods

XI. Software Quality (6–8% questions)
    A. Software Quality Concepts
    B. Planning for SQA and V&V
    C. Methods for SQA and V&V
    D. Measurement Applied to SQA and V&V

[1]http://www.computer.org/certification/Specifications.htm.

This certification examination is not for recent graduates. Individuals applying for CSDP certification must satisfy the educational and experiential requirements. At the time of application, each candidate must have a baccalaureate or equivalent university degree and a minimum of 9,000 hours of recent software engineering experience within at least six (6) of the eleven (11) knowledge areas provided previously.

One benefit that occurs as a by-product of exam preparation is that individuals often increase both their depth and breadth of domain knowledge. CSDP certification holders are also required to continue their education in support of 3-year recertification requirements. CSDP certifications expire within this 3-year period unless the certificate holder can demonstrate that these continuing education requirements have been met.

Set Realistic Goals (Diagnose)

The Software CMM® and CMMI®-Staged provide the building blocks for higher maturity at each stage. It is important to set realistic goals when beginning down the path of process improvement. The leap from chaos (Level 1) to Level 2 is often the hardest step for many organizations. Defining the initial process baseline is key—in order to understand where the organization needs to be, it must first understand where it is (see Table 6-5).

Use the CMM® or CMMI®-SW (Staged) Level 2 KPAs to identify areas of weakness or bottlenecks in existing processes. Then refer to each of the appropriate IEEE software engineering standards using them as planning tools and as checklists to be considered when determining how to accomplish process completeness.

The Software Engineering Institute provides a maturity questionnaire that may be used to provide a snapshot of organizational process maturity. This software process maturity questionnaire is based on the SW-CMM® (see Table 6-6). It has been designed for use is support of CMM®-based software process appraisal methods: the CMM-based appraisal for internal process improvement (CBA IPI) and CMM®-based software capability evaluations (SCEs). This questionnaire is organized by SW-CMM® key process areas (KPAs) and covers all 18 KPAs [60].

The questionnaire described in Table 6-6 provides a high-level look at the process capability for a given organization. It can be given to each participant in under an hour and has the advantage of being able to provide a quick look into the software processes of a given organization. Obvious problem areas or gaps can be easily identified in a relatively short period of time. This method of gap analysis has the disadvantage of not providing the detail that is often required to refine implemented processes.

It is important to identify which organizational process plans will be developed and the sequence of their development. Many organizations begin with the definition of their SCM policies and practices. Beginning with SCM allows organizations to realize tangible, immediate, results. This can provide momentum for the further development of the processes supporting software project planning, software requirements management, software metrics definition, SQA, and processes supporting software verification and validation.

Fix Timelines (Establish)

Define your adoption strategy. This is the strategy used by the organization to facilitate process institutionalization. This strategy should include targeted projects, key personnel,

Table 6-5 Determine if essential processes are missing or are incomplete [72]

Process Management	Engineering
Organizational Process Focus	Requirements Management
Organizational Process Definition	Requirements Development
Organizational Training	Technical Solution
Organizational Process Performance	Product Integration
Organizational Innovation and Deployment	Verification
	Validation
Project Management	Support
Project Planning	Configuration Management
Project Monitoring and Control	Process and Product Quality Assurance
Supplier Agreement Management	Measurement and Analysis
Integrated Product Management	Decision Analysis and Resolution
Risk Management	Organizational Environment for Integration
Integrated Teaming	Causal Analysis and Resolution
Integrated Supplier Management	
Quantitative Project Management	

Table 6-6 Examples from SW-CMM® Maturity Questionnaire

Software Requirements
    Are system requirements allocated to software used to establish a baseline for software engineering and management use?
    As the systems requirements allocated to software change, are the necessary adjustments to software plans, work products, and activities made?

Software Project Planning
    Are estimates (e.g., size, cost, and schedule) documented for use in planning and tracking the software project?
    Do the software plans document the activities to be performed and the commitments made for the software project?

Software Project Tracking and Oversight
    Are the project's actual results (e.g., schedule, size, and cost) compared with estimates in the software plans?
    Is corrective action taken when actual results deviate significantly from the project's software plans?

Software Subcontractor Management
    Is a documented procedure used for selecting subcontractors based on their ability to perform the work?
    Are changes to subcontracts made with the agreement of both the prime contractor and the subcontractor?

Software Quality Assurance
    Are the results of SQA reviews and audits provided to affected groups and individuals (e.g., those who performed the work and those who are responsible for the work)?
    Are issues of noncompliance that are not resolved within the software project addressed by senior management (e.g., deviations from applicable standards)?

Software Configuration Management
    Has the project identified, controlled, and made available the software work products through the use of configuration management?
    Does the project follow a documented procedure to control changes to configuration items/units?

---

training, and, most importantly, schedules reflecting specific software process improvement targets. Goal-driven process improvement is the most effective. Identify both short- and long-term goals, stating concise objectives and time periods, and associate these goals as schedule milestones (see Table 6-7).

Baseline and Implement Processes (Act)

Defining a process baseline is critical when implementing software processes that can be repeatable (see Table 6-8). Use IEEE standards to develop your baseline process documentation that addresses the requirements of each Level 2 KPA. It is also important to evaluate and identify any potential tools that may be used in support of process automation. An ideal candidate area for this type of automation is SCM. There are a number of widely marketed tools that support the documentation and control of various types of software configuration items.

    Take advantage of the information provided by the IEEE Software Engineering Standards Collection. Many of these standards provide documentation templates and describe in detail what individual project support processes should contain. Think of the standards as an in-house software process consultant who has recommended, based upon years of

Table 6-7 Example implementation timeline

(0–3 months)
- Identify individuals responsible for software process improvement.
- Identify project managers who will be participating.
- Identify list of candidate projects.
- Solidify backing of senior management.
- Look at existing processes and mak sure they are appropriate and reflect current business needs (small vs. large projects) using CMM®/CMMI® KPAs and IEEE software engineering standards.
- Define process plans (software configuration management plan, software requirements management plan, software quality assurance plan) using IEEE software engineering standards and "bump" them against the CMM®/CMMI®-SW (Staged) (Level 2).
- Get project members to provide feedback on process plans, review and incorporate feedback.
- Conduct ARC Class C Gap Analysis.

(3–6 months)
- Create process document templates for project documentation based upon defined processes; projects will use these to develop their own plans (e.g., software development plan, software requirements specification).
- Conduct weekly/monthly status reports/reviews to gauge and report progress and provide areas for improvement.

(6–9 months)
- Conduct CMM® Level 2 reviews of the projects. It would be ideal to also include members from unselected projects to participate in these reviews, along with reporting senior management.
- Provide feedback regarding project review, providing requirements for improvement to the projects.

(9–12 months)
- Conduct Level 2 internal assessment, with reporting to senior management.
- Provide feedback regarding project review, providing requirements for improvement to the projects.

Table 6-8 Example action plan

Process Area	Weakness or Area for Improvement	Short Description of How to Address	Project Point of Contact	Resolution Date
Project Montoring and Control	SG2. Although corrective actions are generally tracked to closure, the effectiveness of corrective actions is not consistently addressed.	Create an analysis procedure. This procedure must include a review of the results and effectiveness of corrective actions.	Jim Smith	

experience, the proper methodologies and techniques to be used in support of software development.

Perform Gap Analysis (Learn)

It is important to gauge how effectively process improvements have been implemented for continuous process improvement to be successful. This can be determined through the

Table 6-9 SW-CMM® and CMMI® appraisals

...mework	SW-CMM®	CMMI®
...ramework	CAF	ARC
	CBA-IPI®	SCAMPI
...valuation	SCE	SCAMPI

development of a benchmarking appraisal to support gap analysis activities. This type of appraisal will provide a baseline for future process improvement efforts and will identify weaknesses and strengths.

To begin, review the associated appraisal methodology that is used in support of the SW-CMM® or the CMMI® (see Table 6-9). Use these requirements, in conjunction with the SW-CMM® or the CMMI-Staged (SW) Level 2 KPA criteria, to develop assessment matrixes. These matrixes can be used to determine model compliance and identify specific deficiencies in implemented practices, or the support documentation that was developed using the IEEE software engineering standards set.

Perform Self-Audit Using SW-CMM® KPAs

In order to identify specific weaknesses and associate these to individual practices, an organization should use the key practices of the SW-CMM® to form the basis for a gap analysis. These key practices are described in detail in the SEI technical report, *Key Practices of the Capability Maturity Model* [55]. In this document, each KPA is broken down to all supporting practice commitments, abilities, activities, measurements, and verifications. Supporting information, which provides additional clarification, is also supplied for each practice. The most straightforward way of determining process model compliance is to address each of these practices with an honest answer to the question of whether or not an organization is compliant [49] (see Table 6-10).

Perform Self-Audit Using CMMI®-SW (Staged) KPAs

As previously detailed for the organization using the SW-CMM®, the components of the CMMI®-SW (Staged) can be used to form the basis for a gap analysis. These key components are described in detail in the SEI technical report, *Capability Maturity Model Integration (CMMI), Version 1.1* [57]. In this document, each required, expected, and informative component is addressed in detail. Supporting information is also supplied for each

Table 6-10 Example of SW-CMM® compliance matrix

Req #	Source	Requirement	Satisfied By	Status (P/F/NI)
1	SCM Ab1.4	How does the CCB authorize the creation of products from the baseline library?	Support CCB reviews	Pass, fail, or needs improvement (with comments)
2	SCM Ac 5	How are change requests for all configuration items tracked	Perform change control processing	

Table 6-11 Example of CMMI®-SW (Staged) Compliance Matrix

Req #	Source	Requirement	Satisfied By	Status (P/F/NI)
1	RM SP1.2	How are commitments from participants obtained?	Impact assessments; Documented commitment and requirements changes	Pass, fail, or needs improvement (with comments)
2	RM SP1.3	How are requirements managed as they evolve during the project?	Requirements status reporting; Requirements tracking database	

specific and generic practice. Use this detailed information to identify areas of compliance, noncompliance, or areas needing improvement (see Table 6-11).

IMPLEMENTATION PITFALLS

The implementations of the CMM®/CMMI®-SW (Staged) are fraught with some common pitfalls. It is important to remember that these models are not prescriptive. Too often, organizations implement process control and improvement, spending considerable time and effort, without realizing significant improvements in product cost, quality, or cycle time.

Being Overly Prescriptive

Implementing the recommended practices verbatim can be costly and may not reflect the specific process requirements of an organization. This approach can result in an overly prescriptive process that will increase cost and slow product cycles. This can rapidly destroy the credibility of the process improvement implementation with management and the software development staff. Carefully analyze what areas of process improvement will provide the most positive impact on existing programs; implement these first and management will see the tangible results and continue their support. Small projects may require less formality in planning than large projects, but all components of each standard should be addressed by every software project. Components may be included in the project level documentation, or they may be merged into a system-level or business-level plan, depending upon the complexity of the project.

Remaining Confined to a Specific Stage

Be careful to not examine the implementation of each stage in isolation. There is certainly room to introduce a few higher-maturity-level practices early on if there is a good business case for them. For example, peer reviews are a Level 3 requirement but an organization can gain significant product improvement and insight into their development process with their implementation during Level 2. Also, many organizations that are just beginning a process improvement effort often delay implementation of a measurement program. Without measurements, an organization can achieve maturity Level 3 without understanding the true costs and benefits associated with process improvement.

Also, be aware that projects within the same organization may be at different maturity levels. In a strictest application of the staged representation, the organization would require that projects be at maturity Level 2 before moving any of the projects to level 3. This can take a very long time and is counterproductive. Most organizations identify several pilot projects and move them through the maturity levels. The experience gained from this type of implementation is then carried throughout the entire organization to the remaining efforts.

The CMMI® offers a continuous representation as an alternative to the traditional staged representation, but many organizations are intimidated by the complexity of the models and they are reluctant to make the transition.

Documentation, Documentation

Be careful not to generate policies and procedures to simply satisfy the model CMM®/CMMI® model requirements. Generating documentation for the sake of documentation is a waste of time and resources. Policies and procedures should be developed at the organizational level and used by all projects, only requiring documentation where there is deviation from the standard. The pairing of IEEE standards, which are prescriptive and describe software engineering minimums, with these models can reduce this risk significantly.

Lack of Incentives

The CMM® recommends creating a Software Engineering Process Group (SEPG) chartered with the responsibility of defining and improving the organization's software process. The CMMI® talks about an Engineering Process Group (EPG), reflecting its broader focus. Many times, the success of the process improvement initiatives is placed on the shoulders of these groups. The SEPG and EPG should act as mentors, guiding the software projects through continuous process improvement. Each selected project should be required to follow the recommendations of these groups. Process improvement is most effective and the results more permanent when the projects are stakeholders in the software improvement process.

No Metrics

If your organization has effectively implemented process improvement, you can realize improved business results. However, too often organizations implement process improvement just to meet the process goals for certification purposes. Inefficient implementation will be the result when organizational needs are not placed as a priority. Process performance can be static, or even degrade, relative to business goals. The models don't require substantial process performance measurements until the higher maturity levels. Without a solid understanding of the cost/benefit ratio and its relationship to business results, it can be easy to lose corporate support.

CONCLUSION

Process improvement can be intimidating. Many times, the task of process improvement comes in the form of a directive from senior management, or as a customer requirement, leaving those assigned with a feeling of helplessness. However, all those practicing as software engineers should desire to evolve out of the chaotic activities associated with un-

CONCLUSION **131**

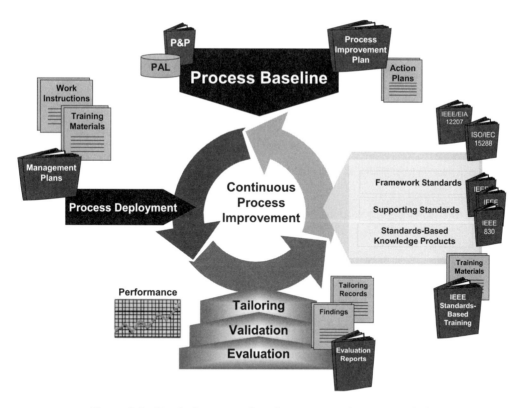

Figure: 6-1 Standards support of continuous process improvement [65].

Table 6-12 Level 2 CMM®/IEEE standards high-level support matrix

Level 2 CMM®	IEEE Standards
Requirements Management	IEEE Std 830-1998
	IEEE Recommended Practice for Software Requirements Specifications
Software Project Planning	IEEE Std 1058-1998
	IEEE Standard for Software Project Management Plans
Software Project Tracking and Oversight	IEEE Std 1058-1998
	IEEE Standard for Software Project Management Plans
Software Quality Assurance	IEEE Std 730-2002
	IEEE Standard for Software Quality Assurance
Software Configuration Management	IEEE Std 828-1998
	IEEE Standard for Software Configuration Management Plans
Software Subcontract Management	IEEE Std 1062-1998
	IEEE Recommended Practice for Software Acquisition

Table 6-13 Level 2 CMMI®-SW (Staged)/IEEE standards comparison matrix

Level 2 CMMI-SW (Staged)	IEEE Standards
Requirements Management	IEEE Std 830-1998
	IEEE Recommended Practice for Software Requirements Specifications
Project Planning	IEEE Std 1058-1998
	IEEE Standard for Software Project Management Plans
Project Monitoring and Control	IEEE Std 1058-1998
	IEEE Standard for Software Project Management Plans
Process and Product Quality Assurance	IEEE Std 730-2002
	IEEE Standard for Software Quality Assurance
Configuration Management	IEEE Std 828-1998
	IEEE Standard for Software Configuration Management Plans
Supplier Agreement Management	IEEE Std 1062-1998
	IEEE Recommended Practice for Software Acquisition
Measurement and Analysis	IEEE Std 1061-1998
	IEEE Standard for a Software Quality Metrics Methodology

controlled software processes, and required heroic efforts, of a Level 1 organization. At the repeatable/managed level, Level 2, software engineering processes are under basic management control and there is an established management discipline; this provides benefits to all involved. The customer may be assured of a lower risk of failure, the organization is provided with accurate insight into the effort, management can more effectively identify and elevate development issues, and team members can work to efficiently managed baselines (see Figure 6-1).

The Level 2 requirements of the SW-CMM® and CMMI®-SW (Staged) are broad. No single IEEE standard can be used in isolation to support these requirements. Rather, a subset of the available IEEE software engineering standards should be employed in combination to provide effective support for Level 2 CMM® activities (see Table 6-11).

IEEE software engineering standards can be used to provide detailed, prescriptive, support for SW-CMM®/CMMI®-SW (Staged) process definition and improvement activities (see Table 6-12).

IEEE standards specify the contents of required documents, the activities that must be performed to fulfill the intent of the standard, and the individuals or groups responsible for these activities. These standards are an extremely valuable source for organizational and project-level process improvement activities and facilitate the development of tailored supporting software processes. Remember to tailor these standards to suit your processes. IEEE standards can be used to form a foundation of effective SEI CMM® Level 2 compliant software engineering practices.

APPENDIX A

IEEE Standards Abstracts

The IEEE Standards collection can be grouped into five topical categories of support: Standards used to provide a common understanding of a domain area, standards that focus on software lifecycle development and definition, standards that concentrate on software engineering process areas, standards that concentrate on the product, and those used to define software engineering techniques.

Standard Number	Standard Name	Description	Comments
Customer and Terminology Standards			
IEEE Std 610.12-1990 (Sept 28), Reaffirmed Sept 2002	IEEE Standard Glossary of Software Engineering Terminology	IEEE Std 610.12-1990, IEEE Standard Glossary of Software Engineering Terminology, identifies terms currently in use in the field of software engineering. Standard definitions for those terms are established.	Provides a common language when documenting software efforts.
IEEE Std 1062-1998 Edition (Dec 2), Reaffirmed Sept 2002	IEEE Recommended Practice for Software Acquisition	A set of useful quality practices that can be selected and applied during one or more steps in a software acquisition process is described. This recommended practice can be applied to software that runs on any computer system regardless of the size, complexity, or criticality of the software, but is more suited for use on modified-off-the-shelf software and fully developed software.	Good support document for individuals or organizations that use software and acquire that software from suppliers.

Standard Number	Standard Name	Description	Comments
IEEE Std 1220-1998 (Dec 8)	IEEE Standard for the Application and Management of the Systems Engineering Process	The interdisciplinary tasks, which are required throughout a system's life cycle to transform customer needs, requirements, and constraints into a system solution, are defined. In addition, the requirements for the systems engineering process and its application throughout the product life cycle are specified. The focus of this standard is on engineering activities necessary to guide product development while ensuring that the product is properly designed to make it affordable to produce, own, operate, maintain, and eventually to dispose of, without undue risk to health or the environment.	High-level documentation for those who are responsible for the oversight of Software Engineering processes.
IEEE Std 1228-1994 (Mar 17), Reaffirmed Dec 2002	IEEE Standard for Software Safety Plans	The minimum acceptable requirements for the content of a software safety plan are established. This standard applies to the software safety plan used for the development, procurement, maintenance, and retirement of safety-critical software. This standard requires that the plan be prepared within the context of the system safety program. Only the safety aspects of the software are included. This standard does not contain special provisions required for software used in distributed systems or in parallel processors.	Provides support to those involved with safety-critical software in the development of safety plans that address the potential software safety risks.
IEEE Std 1233, 1998 Edition (Apr 17), Reaffirmed Sept 2002	IEEE Guide for Developing System Requirements Specifications	Guidance for the development of the set of requirements, system requirements specification (SyRS), that will satisfy an expressed need is provided. Developing a SyRS includes the identification, organization, presentation, and modification of the requirements. Also addressed are the conditions for incorporating operational concepts, design constraints, and design configuration requirements into the specification. This guide also covers the necessary characteristics and qualities of individual requirements and the set of all requirements.	Good reference material, used by requirements analysts to develop an SRS and define requirements. It should be used to clarify what constitutes a good requirement and provide an understanding of where to look to identify different requirement sources.

Standard Number	Standard Name	Description	Comments
IEEE Std 1362-1998 (Mar 19)	IEEE Guide for Information Technology—System Definition—Concept of Operations (ConOps) Document	The format and contents of a concept of operations (ConOps) document are described. A ConOps is a user-oriented document that describes system characteristics for a proposed system from the users' viewpoint. The ConOps document is used to communicate overall quantitative and qualitative system characteristics to the user, buyer, developer, and other organizational elements (for example, training, facilities, staffing, and maintenance). It is used to describe the user organization(s), mission(s), and organizational objectives from an integrated systems point of view.	Use this document for support if required to write a concept of operations.
IEEE Std 1517-1999 (Jun 26)	IEEE Standard for Information Technology—Software Life Cycle Processes—Reuse Processes	A common framework for extending the software life cycle processes of IEEE/EIA Std 12207.0-1996 to include the systematic practice of software reuse is provided. This standard specifies the processes, activities, and tasks to be applied during each phase of the software life cycle to enable a software product to be constructed from reusable assets. It also specifies the processes, activities, and tasks to enable the identification, construction, maintenance, and management of assets supplied.	Describes a high-level framework for reuse processes, but not the details of how to perform the activities and tasks included in the processes
IEEE Std 1540-2001 (Mar 17)	IEEE Standard for Software Life Cycle Processes—Risk Management	A process for the management of risk in the life cycle of software is defined. It can be added to the existing set of software life cycle processes defined by the IEEE/EIA 12207 series of standards, or it can be used independently.	This standard establishes minimum requirements for software risk management process, activities, and tasks.
IEEE Std 1471-2000 (Sept 21)	IEEE Recommended Practice for Architectural Description of Software Intensive Systems	This recommended practice addresses the activities of the creation, analysis, and sustainment of architectural descriptions. A conceptual framework for architectural description is established. The content of an architectural description is defined. Annexes provide the rationale for key concepts and terminology, the relationships to other standards, and examples of usage.	Great for individuals who develop, describe, maintain and document architectures (architects). Also useful for those who oversee and evaluate systems and their development.

Standard Number	Standard Name	Description	Comments
Life Cycle Standards			
IEEE/EIA 12207.0-1996 (Mar)	Standard for Information Technology—Software Life Cycle Processes	ISO/IEC 12207 provides a common framework for developing and managing software. IEEE/EIA 12207.0 consists of the clarifications, additions, and changes accepted by the Institute of Electrical and Electronics Engineers (IEEE) and the Electronic Industries Association (EIA) as formulated by a joint project of the two organizations.	This standard establishes a common framework for software life cycle processes, with well-defined terminology that can be referenced by the software industry.
IEEE/EIA 12207.1-1996 (April)	Standard for Information Technology—Software Life Cycle Processes—Life Cycle Data	ISO/IEC 12207 provides a common framework for developing and managing software. IEEE/EIA 12207.0 consists of the clarifications, additions, and changes accepted by the Institute of Electrical and Electronics Engineers (IEEE) and the Electronic Industries Association (EIA) as formulated by a joint project of the two organizations. IEEE/EIA 12207.1 provides guidance for recording life cycle data resulting from the life cycle processes of IEEE/EIA 12207.0.	To be used with IEEE/EIA 12207.0 and IEEE/EIA 12207.2; describes basic and supporting life cycle process data. This guide defines the life cycle data of IEEE/EIA 12207.0 by relating the tasks and activities defined in IEEE/EIA 12207.0 with the following kinds of documentation: description, plan, procedure, record, report, request, and specification.
IEEE/EIA 12207.2-1997 (Apr 1998)	Standard for Information Technology—Software Life Cycle Processes—Implementation Considerations	ISO/IEC 12207 provides a common framework for developing and managing software. IEEE/EIA 12207.0 consists of the clarifications, additions, and changes accepted by the Institute of Electrical and Electronics Engineers (IEEE) and the Electronic Industries Association (EIA) as formulated by a joint project of the two organizations. IEEE/EIA 12207.2 provides implementation consideration guidance for the normative clauses of IEEE/EIA 12207.0. The guidance is based on software industry experience with the life cycle processes presented in IEEE/EIA 12207.0.	To be used with IEEE/EIA 12207.0 and IEEE/EIA 12207.1; describes implementation considerations for lifecycle processes. This guide provides implementation consideration guidance for the normative clauses of IEEE/EIA 12207.0 that is based on industry experience with the life cycle processes presented in IEEE/EIA 12207.0.

Standard Number	Standard Name	Description	Comments
IEEE Std 14143.1-2000 (Jan 30)	IT—Software Measurement—Functional Size Measurement. Part 1: Definition of Concepts	Implementation notes that relate to the IEEE interpretation of ISO/IEC 14143-1:1998 are described.	IEEE Std 14143.1-2000 deals with the fundamental concepts that apply to the general class of functional size measurement (FSM) methods.

Process Standards

Standard Number	Standard Name	Description	Comments
IEEE Std 730-2002 (Oct 20), Revised Sept 2002	IEEE Standard for Software Quality Assurance Plans	Uniform, minimum acceptable requirements for preparation and content of software quality assurance plans (SQAPs) are provided. This standard applies to the development and maintenance of critical software. For noncritical software, or for software already developed, a subset of the requirements of this standard may be applied.	A great reference for SQA plan development. Describes the minimum acceptable requirements for preparation and content of software quality assurance plans.
IEEE Std 828-1998 (Jun 25)	IEEE Standard for Software Configuration Management Plans	The minimum required contents of a software configuration management plan (SCMP) are established, and the specific activities to be addressed and their requirements for any portion of a software product's life cycle are defined.	A valuable SCM Plan development reference. Describes the minimum acceptable requirements for preparation and content of software configuration management plans.
ANSI/IEEE Std 1008-1987 (R1993), Reaffirmed Dec 2002	An American National Standard—IEEE Standard for Software Unit Testing	This standard's primary objective is to specify a standard approach to software unit testing that can be used as a basis for sound software engineering practice.	Describes best practice and formalizes the "desk check."
IEEE Std 1012-1998 (Mar 9)	IEEE Standard for Software Verification and Validation	Software verification and validation (V&V) processes, which determine whether development products of a given activity conform to the requirements of that activity, and whether the software satisfies its intended use and user needs, are described. This determination may include analysis, evaluation, review, inspection, assessment, and testing of software products and processes. V&V processes assess the software in the context of the system, including the operational environment, hardware, interfacing software, operators, and users.	Establishes a common framework for processes, activities, and tasks in support of all software life cycle processes, including acquisition, supply, development, operation, and maintenance processes. Define the content of a software V&V plan.

Standard Number	Standard Name	Description	Comments
IEEE Std 1012a-1998 (Sept 16)	Supplement to IEEE Standard for Software Verification and Validation: Content Map to IEEE/EIA 12207.1-1996	The relationship between the two sets of requirements on plans for verification and validation of software, found in IEEE Std 1012-1998 and IEEE/EIA 12207.1-1996, is explained so that users may produce documents that comply with both standards.	Provides a bridge between two plans that are complex and overlap.
IEEE Std 1028-1997 (Mar 4), Reaffirmed Sept 2002	IEEE Standard for Software Reviews	This standard defines five types of software reviews, together with procedures required for the execution of each review type. This standard is concerned only with the reviews; it does not define procedures for determining the necessity of a review, nor does it specify the disposition of the results of the review. Review types include management reviews, technical reviews, inspections, walkthroughs, and audits.	Useful when there is a requirement to define formal reviews and procedures.
IEEE Std 1042-1987. Not maintained but available	IEEE Guide to Software Configuration Management	This guide describes the application of configuration management disciplines to management of software engineering projects. This guide serves three groups: developers of software, the software management community, and those responsible for preparation of SCM plans. Software configuration management consists of two major aspects: planning and implementation. This guide focuses on software configuration management planning and provides broad perspectives for the understanding of software configuration management.	Provides a good foundation for software configuration management training.
IEEE Std 1045-1992 (Sep 17), Reaffirmed Dec 2002	IEEE Standard for Software Productivity Metrics	Consistent ways to measure the elements that go into computing software productivity are defined. Software productivity metrics terminology is given to ensure an understanding of measurement data for both source code and document production.	Very complex but provides useful information when required to measure productivity. The goal of this standard is a better understanding of the software process, which may lend insight to improving it.
IEEE Std 1058-1998 (Dec 8)	IEEE Standard for Software Project Management Plans	The format and contents of software project management plans, applicable to any type or size of software project, are described. The elements that should appear in all software project management plans are identified.	Great for project managers; provides a description of all PM areas of concern.

Standard Number	Standard Name	Description	Comments
IEEE Std 1074-1997 (Dec 9)	IEEE Standard for Developing Software Life Cycle Processes	A process for creating a software life cycle process is provided. Although this standard is directed primarily at the process architect, it is useful to any organization that is responsible for managing and performing software projects.	Good reference material for life cycle process creation.
IEEE Std 1219-1998 (Jun 25)	IEEE Standard for Software Maintenance	The process for managing and executing software maintenance activities is described.	Good for helping to define software maintenance activities
IEEE Std 1490-1998 (Jun 25)	IEEE Guide—Adoption of PMI Standard—A Guide to the Project Management Body of Knowledge	The subset of the Project Management Body of Knowledge that is generally accepted is identified and described in this guide. "Generally accepted" means that the knowledge and practices described are applicable to most projects most of the time, and that there is widespread consensus about their value and usefulness. It does not mean that the knowledge and practices should be applied uniformly to all projects without considering whether they are appropriate.	Adds value when combined with 1058.

Resource and Technique Standards

IEEE Std 829-1998 (Sept 16)	IEEE Standard for Software Test Documentation	A set of basic software test documents is described. This standard specifies the form and content of individual test documents. It does not specify the required set of test documents.	A good source for testing documentation. Describes the minimum acceptable requirements for documentation in support of software testing.
IEEE Std 830-1998 (Jun 25)	IEEE Recommended Practice for Software Requirements Specifications	The content and qualities of a good software requirements specification (SRS) are described and several sample SRS outlines are presented. This recommended practice is aimed at specifying requirements of software to be developed but also can be applied to assist in the selection of in-house and commercial software products.	A great starting point for creating an SRS. Guidelines for compliance with IEEE/EIA 12207.1-1996 are also provided.
IEEE Std 1016-1998 (Sept 23)	IEEE Recommended Practice for Software Design Descriptions	The necessary information content and recommendations for an organization for software design descriptions (SDDs) are described. An SDD is a representation of a software system that is used as a medium for communicating software design information. This recommended practice is applicable to paper documents, automated databases, design description languages, or other means of description.	This is a great tool for the support of design documentation. It specifies the required information content and recommended organization for a software design description (SDD). The practice may be

Standard Number	Standard Name	Description	Comments
IEEE Std 1016-1998 (Sept 23) (*cont.*)			applied to all application domains and is not restricted by the size, complexity, or criticality of the software.
IEEE Std 1044-1993 (Dec 2), Reaffirmed Sept 2002	IEEE Standard Classification for Software Anomalies	A uniform approach to the classification of anomalies found in software and their documentation is provided. The processing of anomalies discovered during any software life cycle phase are described, and comprehensive lists of software anomaly classifications and related data items that are helpful to identify and track anomalies are provided.	This standard is not intended to define procedural or format requirements for using the classification scheme. It does, however, identify some classification measures.
IEEE Std 1175-1991 (Dec 5)	IEEE Standard Reference Model for Computing System Tool Interconnections	The purpose is to establish agreements for information transfer among tools in the contexts of human organization, a computer system platform, and a software development application .Interconnections that must be considered when buying, building, testing, or using computing system tools for specifying behavioral descriptions or requirements of system and software products are described.	Makes a simple concept hard to understand. Reference models for tool-to-organization interconnections, tool-to- platform interconnections, and information transfer among tools are provided
IEEE Std 1175.1-2002 (Nov. 11)	IEEE Guide for CASoftware Engineering Tool Interconnections— Classification and Description	Introduces and characterizes the problem of interconnecting CASoftware Engineering tools with their environment.	This guide is intended to help buyers, builders, testers, users, and managers responsible for the implementation of computing system tools reach sound decisions on an implementation strategy.
IEEE Std 1320.1-1998 (Jun 25)	IEEE Standard for Functional Modeling Language— Syntax and Semantics for IDEF0	IDEF0 function modeling is designed to represent the decisions, actions, and activities of an existing or prospective organization or system. IDEF0 may be used to model a wide variety of systems composed of people, machines, materials, computers, and information of all varieties and structured by the relationships among them, both automated and nonautomated. As the basis of this architecture, IDEF0 may then be used to design an implementation that meets these requirements and performs these functions.	IDEF0 graphics and accompanying texts are presented in an organized and systematic way to gain understanding, support analysis, provide logic for potential changes, specify requirements, and support system--level design and integration activities. Good supplement for design and integration activities.

Standard Number	Standard Name	Description	Comments
IEEE Std 1320.2-1998 (Jun 25). IEEE Std 1320.2a	IEEE Standard for Conceptual Modeling Language Syntax and Semantics for IDEF1X 97 (IDEF object)	IDEF1X 97 consists of two conceptual modeling languages. The key-style language supports data/information modeling and is downward compatible with the U.S. government's 1993 standard, FIPS PUB 184. The identity-style language is based on the object model with declarative rules and constraints.	A great starting point for database modeling. IDEF1X 97 conceptual modeling supports implementation by relational databases, extended relational databases, object databases, and object programming languages.
IEEE Std 1420.1-1995 (Dec 12), Reaffirmed June 2002	IEEE Standard for IT Software Reuse—Data Model for Reuse Library Interoperability: Basic Interoperability Data Model (BIDM)	The minimal set of information about assets that reuse libraries should be able to exchange to support interoperability is provided.	Reuse reference material. This document describes the Basic Interoperability Data Model (BIDM) standard as developed by the Reuse Library Interoperability Group (RIG).
IEEE Std 1420.1a-1996 (Dec 10), Reaffirmed June 2002	Supplement to IEEE Standard for IT Software Reuse—Data Model for Reuse Library Interoperability: Asset Certification Framework	A consistent structure for describing a reuse library's asset certification policy in terms of an asset certification framework is defined, along with a standard interoperability data model for interchange of asset certification information.	This document defines a consistent structure for describing a reuse library's asset certification policy in terms of an asset certification framework.
IEEE Std 1420.1b-1999 (Jun 26), Reaffirmed June 2002	IEEE Trial-Use Supplement to IEEE Standard for IT Software Reuse—Data Model for Reuse Library Interoperability: Intellectual property Rights Framework	This extension to the Basic Interoperability Data Model (IEEE Std 1420.1-1995) incorporates intellectual property rights issues into software asset descriptions for reuse library interoperability.	The intellectual property rights framework defines a standard for the consistent structure, labeling, and description of intellectual property rights management policies and procedures for reuse assets.

Standard Number	Standard Name	Description	Comments
IEEE Std 1462-1998 (Mar 19)	IEEE Standard—Adoption of ISO/IEC 14102: 1995— Information Technology—Guideline for the evaluation and selection of CASoftware Engineering tools	ISO/IEC 14102:1995 deals with the evaluation and selection of CASoftware Engineering tools, covering a partial or full portion of the software engineering life cycle. The adoption of the international standard by IEEE includes an implementation note, which explains terminology differences, identifies related IEEE standards, and provides interpretation of the international standard.	Focus on the evaluation and selection of CASoftware Engineering tools establishing processes and activities to be applied for the evaluation of CASoftware Engineering tools. Processes are generic and must be tailored to meet specific organizational requirements.
IEEE Std 2001-2002 (Jan 21, 2003)	IEEE Recommended Practice for Internet Practices—Web Page Engineering—Intranet/Extranet Applications	This standard defines recommended practices for Web page engineering. It addresses the needs of Webmasters and managers to effectively develop and manage World Wide Web projects (internally via an intranet or in relation to specific communities via an extranet).	This standard discusses life cycle planning—identifying the audience, the client environment, objectives, and metrics—and continues with recommendations on server considerations and specific Web page content.

Product Standards

IEEE Std 982.1-1988 (Jun 9)	IEEE Standard Dictionary of Measures to Produce Reliable Software	This standard provides a set of measures indicative of software reliability that can be applied to the software product as well as to the development and support processes. It was motivated by the needs of software developers and users who are confronted with a plethora of models, techniques, and measures.	Provides a common language for software measurement. This standard provides a common, consistent definition of a set of measures.
IEEE Std 1061-1998 (Dec 8)	IEEE Standard for a Software Quality Metrics Methodology	A methodology for establishing quality requirements and identifying, implementing, analyzing, and validating the process and product software quality metrics is defined. The methodology spans the entire software life cycle.	Provides insight into metrics collection, very targeted to quality metrics.
IEEE Std 1063-2001 (Dec 5)	IEEE Standard for Software User Documentation	Two factors motivated the development of this standard: The concern of the software user communities over the poor quality of much user documentation, and a need for requirements expressed by producers of documentation.	Good when developing a software users manual.

Standard Number	Standard Name	Description	Comments
IEEE Std 1465-1998 (Jun 25)	IEEE Standard —Adoption of International Standard ISO/IEC 12119: 1994(E)— Information Technology— Software Packages— Quality Requirements and Testing	Quality requirements for software packages and instructions on how to test a software package against these requirements are established. The requirements apply to software packages as they are offered and delivered, not to the production process (including activities and intermediate products, such as specifications).	May be used to specify quality requirements for software, and provide instructions on how to test against these requirements.

APPENDIX B

Level 2 Mappings of CMMI-SE/SW/IPPD® (Staged) V.1.1 to SW-CMM® V. 1.1

The information provided in the following table was produced by the USAF Software Technology Support Center. [90]

Alphabetical by Abbreviation	Order of Occurrence by Maturity Level
DP—Defect Prevention	RM—Requirements Management
IC—Intergroup Coordination	SPP—Software Project Planning
ISM—Integrated Software Management	SPT&O—Software Project Tracking and Oversight
OPD—Organizational Process Definition	SSM—Software Subcontract Management
OPF—Organizational Process Focus	SQA—Software Quality Assurance
PCM—Process Change Management	SCM—Software Configuration Management
PR—Peer Reviews	
QPM—Quantitative Process Management	
RM—Requirements Management	
SCM—Software Configuration Management	
SPE—Software Product Engineering	
SPP—Software Project Planning	
SPT&O—Software Project Tracking and Oversight	
SQA—Software Quality Assurance	
SQM—Software Quality Management	
SSM—Software Subcontract Management	
TCM—Technology Change Management	
TP—Training Program	

Process Maturity Level 2	CMMI® Process Area	CMMI® Goal	CMMI® Specific Practice or Generic Practice	SW-CMM® V1.1 Goal/Common Feature	Comments
	Requirements Management	SG1	Requirements are managed and inconsistencies with project plans and work products are identified.	RM Goals 1,2	
			SP 1.1. Develop an understanding with the requirements providers on the meaning of the requirements.	IC Ac 1 SPE Ac 2	See subpractice 10
			SP 1.2. Obtain commitment to the requirements from the project participants.	IC Goal 1 RM Ac 1,3 SPE Ac 2	
			SP 1.3. Manage changes to the requirements as they evolve during the project.	RM Ac 3 SCM Ac 5 SPE Ac 2,10 SPT&O Ac 2	
			SP 1.4. Maintain bi-directional traceability among the requirements and the project plans and work products.	SPE Ac 10	Subpractice 3 elevated to specific practice
			SP 1.5. Identify inconsistencies between the project plans and work products and the requirements.	RM Ac 3 SPE Ac 10	
		GG 2	The process is institutionalized as a managed process.		Implied by Level 2
			GP 2.1. Establish and maintain an organizational policy for planning and performing the requirements management process.	RM Co 1	
			GP 2.2. Establish and maintain the plan for performing the requirements management process.		Not directly addressed

Process Maturity Level 2	CMMI® Process Area	CMMI® Goal	CMMI® Specific Practice or Generic Practice	SW-CMM® V1.1 Goal/Common Feature	Comments
			GP 2.3. Provide adequate resources for performing the process, developing the work products, and providing the services of the requirements management process.	RM Ab 3	
			GP 2.4. Assign responsibility and authority for performing the process, developing the work products, and providing the services of the requirements management process.	RM Ab 1	
			GP 2.5. Train the people performing or supporting the requirements management process as needed.	RM Ab 4	
			GP 2.6. Place designated work products of the requirements management process under appropriate levels of configuration management.	SCM Goal 2	
			GP 2.7. Identify and involve the relevant stakeholders of the requirements management process as planned.		Not directly addressed (See GP 2.2 above)
			GP 2.8. Monitor and control the requirements management process against the plan for performing the process and take appropriate corrective action.	RM Me 1, Ve 2	
			GP 2.9. Objectively evaluate adherence of the requirements management process against its process description standards and procedures, and address noncompliance.	RM Ve 3	

Process Maturity Level 2	CMMI® Process Area	CMMI® Goal	CMMI® Specific Practice or Generic Practice	SW-CMM® V1.1 Goal/Common Feature	Comments
			GP 2.10. Review the activities, status, and results of the requirements management process with higher-level management and resolve issues.	RM Ve 1	
	Project Planning	SG1	Estimates of project planning parameters are established and maintained.	SPP Goal 1	
			SP 1.1. Establish a top-level work breakdown structure (WBS) to estimate of the scope of the project.	SPP Ac 5	
			SP 1.2. Establish and maintain estimates of the attributes of the work products and tasks.	SPP Ac 7,9,10	
			SP 1.3. Define the project life-cycle phases upon which to scope the planning effort.	SPP Ac 5,7	
			SP 1.4 Estimate the project effort and cost for the attributes of the work products and tasks based on estimation rationale.	SPP Ac 7,9,10,14	
		SG2	A project plan is established and maintained as the basis for managing the project.	SPP Goal 2 SPP Ac 6,7	
			SP 2.1. Establish and maintain the project's budget and schedule.	SPP Ac 7,12	
			SP 2.2. Identify and analyze project risks.	SPP Ac 7,13	See Ac 7 Subpractice 9
			SP 2.3. Plan for the management of project data.	QPM Ac 1,2,3 SPP Ac 7,8 SPT&O Ac 5,6,7,8,9,10,11 by implication	
			SP 2.4. Plan for necessary resources to perform the project.	SPE Ab 1 SPP Ac 7,11,14	

Process Maturity Level 2	CMMI® Process Area	CMMI® Goal	CMMI® Specific Practice or Generic Practice	SW-CMM® V1.1 Goal/Common Feature	Comments
			SP 2.5. Plan for knowledge and skills needed to perform the project.	ISM Ac 4 SPP Ac 7 TP Ac 1	
			SP 2.6. Plan the involvement of identified stakeholders.	SPP Ac 1,3,6 SPTO Ab 1	
			SP 2.7. Establish and maintain the overall project plan content.	ISM Ac 3 SPP Ac 7 SPT&O Ab 1, Ac 2	
		SG3	Commitments to the project plan are established and maintained.	IC Goal 2 SPP Goal 2,3	
			SP 3.1. Review all plans that affect the project to understand project commitments.	DP Ac 1 PCM Ac 3 QPM Ac 1 SCM Ac 1,2 SPP Ac 3,4,6 SQA Ac 1 SQM Ac 1 SSM Ac 1 TCM Ac 1 TP Ac 1	Informational materials in each usually direct review by the project
			SP 3.2. Reconcile the project plan to reflect available and estimated resources.	SPP Ac 1,4,6, 12,14	Not directly addressed
			SP 3.3. Obtain commitment from relevant stakeholders responsible for performing and supporting plan execution.	IC Ac 3,4,6 SPP Ac 6	
		GG 2	The process is institutionalized as a managed process.		Implied by Level 2
			GP 2.1. Establish and maintain an organizational policy for planning and performing the project planning process.	SPP Co 2	

Process Maturity Level 2	CMMI® Process Area	CMMI® Goal	CMMI® Specific Practice or Generic Practice	SW-CMM® V1.1 Goal/Common Feature	Comments
			GP 2.2. Establish and maintain the plan for performing the project planning process.	SPP Co 2 SPP Ac 6,7,9,10, 11,12 SPT&O Ab1, Ac 2	SW-CMM® v1.1 doesn't always specify, "maintain." Addressed inconsistently in SW-CMM® v1.1
			GP 2.3. Provide adequate resources for performing the project planning process, developing the work products, and providing the services of the process.	SPP Ab 3	
			GP 2.4. Assign responsibility and authority for performing the process, developing the work products, and providing the services of the project planning process	SPP Ab 2 SPP Co 1	
			GP 2.5. Train the people performing or supporting the project planning process as needed.	SPP Ab 4	
			GP 2.6. Place designated work products of the project planning process under appropriate levels of configuration management.	SCM Goal 2	
			GP 2.7. Identify and involve the relevant stakeholders of the project planning process as planned.	SPP Ac 1,3,6	
			GP 2.8. Monitor and control the project planning process against the plan for performing the process and take appropriate corrective action.	SPP Me 1, Ve 2	

Process Maturity Level 2	CMMI® Process Area	CMMI® Goal	CMMI® Specific Practice or Generic Practice	SW-CMM® V1.1 Goal/Common Feature	Comments
			GP 2.9. Objectively evaluate adherence of the project planning process against its process description, standards, and procedures, and address noncompliance.	SPP Ve 3	
			GP 2.10. Review the activities, status, and results of the project planning process with higher-level management and resolve issues.	SPP Ve 1	
	Project Monitoring and Control	SG1	Actual performance and progress of the project is monitored against the project plan.	SG 1 SPT&O Goal 1	
			SP 1.1. Monitor the actual values of the project planning parameters against the project plan.	ISM Ac 6 SPT&O Ac 1,5,6, 7,8,9	
			SP 1.2. Monitor commitments against those identified in the project plan.	SPT&O Ac 8,12	
			SP 1.3. Monitor risks against those identified in the project plan.	ISM Ac 10 SPT&O Ac 10	
			SP 1.4. Monitor the management of project data against the project plan.	SPT&O Ac 11 SPT&O Ve 3	
			SP 1.5. Monitor stakeholder involvement against the project plan.	ISM Ac 9,11 SPT&O Ac 12,13	
			SP 1.6. Periodically review the project's progress, performance, and issues.	ISM Ac 11 SPT&O Ac 4,6,8, 9,12	
			SP 1.7. Review the accomplishments and results of the project at selected project milestones.	SPT&O Ac 12,13	

Process Maturity Level 2	CMMI® Process Area	CMMI® Goal	CMMI® Specific Practice or Generic Practice	SW-CMM® V1.1 Goal/Common Feature	Comments
		SG2	Corrective actions are managed to closure when the project's performance or results deviate significantly from the plan.	SPT&O Goal 2	SW CMM® more rigorous
			SP 2.1. Collect and analyze the issues and determine the corrective actions necessary to address the issues.	SPT&O Ac 5,6,7,8,9	
			SP 2.2. Take corrective action on identified issues.	SPT&O Ac 5,6,7,8,9	
			SP 2.3. Manage corrective actions to closure.	SPT&O Ac 9	
			GP 2.1. Establish and maintain an organizational policy for planning and performing the project monitoring and control process.	SPT&O Co 2	
			GP 2.2. Establish and maintain the plan for performing the project monitoring and control process.	SPT&O Ab 1, Ac 1,2	SW-CMM® v1.1 doesn't directly address a plan for performing the SPT&O process.
			GP 2.3. Provide adequate resources for performing the project monitoring and control process, developing the work products, and providing the services of the process.	SPT&O Ab 3	
			GP 2.4. Assign responsibility and authority for performing the process, developing the work products, and providing the services of the project monitoring and control process.	SPT&O Ab 2, Co 1	

Process Maturity Level 2	CMMI® Process Area	CMMI® Goal	CMMI® Specific Practice or Generic Practice	SW-CMM® V1.1 Goal/Common Feature	Comments
			GP 2.5. Train the people performing or supporting the project monitoring and control process as needed.	SPT&O Ab 4,5	
			GP 2.6. Place designated work products of the project monitoring and control process under appropriate levels of configuration management.	SCM Goal 2	
			GP 2.7. Identify and involve the relevant stakeholders of the project monitoring and control process as planned.	ISM Ac 9,11 SPT&O Ac 12,13	
			GP 2.8. Monitor and control the project monitoring and control process against the plan for performing the process and take appropriate corrective action.	SPT&O Me 1 SPT&O Ve 2	
			GP 2.9. Objectively evaluate adherence of the project monitoring and control process against its process description, standards and procedures, and address noncompliance.	SPT&O Ve 3	
			GP 2.10. Review the activities, status, and results of the project monitoring and control process with higher-level management and resolve issues.	SPT&O Ve 1	
	Supplier Agreement Management	SG1	Agreements with the suppliers are established and maintained.	SSM Goal 2,3	

Process Maturity Level 2	CMMI® Process Area	CMMI® Goal	CMMI® Specific Practice or Generic Practice	SW-CMM® V1.1 Goal/Common Feature	Comments
			SP 1.1. Determine the type of acquisition for each product or product component to be acquired.		Not addressed
			SP 1.2. Select suppliers based on an evaluation of their ability to meet the specified requirements and established criteria.	SSM Ac 2 SSM Goal 1	
			SP 1.3. Establish and maintain formal agreements with the supplier.	SSM Ac 6	
		SG2	Agreements with the suppliers are satisfied by both the project and the supplier.	SSM Ac 3,8	
			SP 2.1. Review candidate COTS products to ensure they satisfy the specified requirements that are covered under a supplier agreement.		Not directly addressed
			SP 2.2. Perform activities with the supplier as specified in the supplier agreement.	SSM AC 3,7,8, 9,13	
			SP 2.3. Ensure that the supplier agreement is satisfied before accepting the acquired product.	SSM Ac 12	
			SP 2.4. Transition the acquired products from the supplier to the project.		Not addressed
		GG2	The process is institutionalized as a managed process.	Implied by Level 2	
			GP 2.1. Establish and maintain an organizational policy for planning and performing the supplier agreement management process.	SSM Co 1	

Process Maturity Level 2	CMMI® Process Area	CMMI® Goal	CMMI® Specific Practice or Generic Practice	SW-CMM® V1.1 Goal/Common Feature	Comments
			GP 2.2. Establish and maintain the plan for performing the supplier agreement management process.	SSM Co 1 SSM Ac 1, 6	SW-CMM® v1.1 doesn't always specify "maintain"
			GP 2.3. Provide adequate resources for performing the supplier agreement management process, developing the work products, and providing the services of the process.	SSM Ab 1	
			GP 2.4. Assign responsibility and authority for performing the process, developing the work products, and providing the services of the supplier agreement management process.	SSM Co 2	
			GP 2.5. Train the people performing or supporting the supplier agreement management process as needed.	SSM Ab 2,3	
			GP 2.6. Place designated work products of the supplier agreement management process under appropriate levels of configuration management.	SCM Goal 2	
			GP 2.7. Identify and involve the relevant stakeholders of the supplier agreement management process as planned.	SSM Ac 1,3,7,8,9	
			GP 2.8. Monitor and control the supplier agreement management process against the plan for performing the process and take appropriate corrective action.	SSM Me 1,Ve 2	

Process Maturity Level 2	CMMI® Process Area	CMMI® Goal	CMMI® Specific Practice or Generic Practice	SW-CMM® V1.1 Goal/Common Feature	Comments
			GP 2.9. Objectively evaluate adherence of the supplier agreement management process against its process description, standards, and procedures, and address noncompliance.	SSM Ve 3	
			GP 2.10. Review the activities, status, and results of the supplier agreement management process with higher-level management and resolve issues.	SSM Ve 1	
	Measurement and Analysis	SG1	Measurement objectives and practices are aligned with identified information needs and objectives.		Not directly addressed
			SP 1.1. Establish and maintain measurement objectives that are derived from identified information needs and objectives.	QPM Co 2, Ac 1 (implied) SPT&O Ac 5,6, 7,8,9,11 (implied)	SW CMM® less rigorous
			SP 1.2. Specify measures to address the measurement objectives.	QPM Ac 3 Implied	
			SP 1.3. Specify how measurement data will be obtained and stored.	QPM Ac 3,4	
			SP 1.4. Specify how measurement data will be analyzed and reported.	QPM Ac 3,5,6 SPT&O Ac 11	
		SG 2.	Measurement results that address identified information needs and objectives are provided.	TCM Ab 4	
			SP 2.1. Obtain specified measurement data.	QPM Ac 4	

Process Maturity Level 2	CMMI® Process Area	CMMI® Goal	CMMI® Specific Practice or Generic Practice	SW-CMM® V1.1 Goal/Common Feature	Comments
			SP 2.2. Analyze and interpret measurement data.	QPM Ac 5 See subpractice 2	
			SP 2.3. Manage and store measurement data, measurement specifications, and analysis results.	OPD Ac 5 QPM Ac 4 SPP Ac 15 SPT&O Ac 11	
			SP 2.4. Report results of measurement and analysis activities to all relevant stakeholders.	QPM Ac 6	
			GG 2. The process is institutionalized as a managed process.	Implied by Level 2	
			GP 2.1. Establish and maintain an organizational policy for planning and performing the measurement and analysis process.	QPM Co 1	Not directly addressed
			GP 2.2. Establish and maintain the plan for performing the measurement and analysis process.	OPF Ac 2 QPM Ac 1 SQM Ac 1,2	Not directly addressed
			GP 2.3. Provide adequate resources for performing the measurement and analysis process, developing the work products, and providing the services of the process.	OPF Ab 2 QPM Ab2,3 SQM Ab 1	Not directly addressed
			GP 2.4. Assign responsibility and authority for performing the process, developing the work products, and providing the services of the measurement and analysis process.		Not directly addressed
			GP 2.5. Train the people performing or supporting the measurement and analysis process as needed.	OPF Ab 3 QPM Ab4 SQM Ab 2,3	Not directly addressed

Process Maturity Level 2	CMMI® Process Area	CMMI® Goal	CMMI® Specific Practice or Generic Practice	SW-CMM® V1.1 Goal/Common Feature	Comments
			GP 2.6. Place designated work products of the measurement and analysis process under appropriate levels of configuration management.	SCM Goal 2	
			GP 2.7. Identify and involve the relevant stakeholders of the measurement and analysis process as planned.	SQM Ac 1	Not directly addressed
			GP 2.8. Monitor and control the measurement and analysis process against the plan for performing the process and take appropriate corrective action.	Most KPAs Ve 2	Not directly addressed
			GP 2.9. Objectively evaluate adherence of the measurement and analysis process against its process description, standards, and procedures, and address non-compliance.	Most KPAs Ve 3	Not directly addressed
			GP 2.10. Review the activities, status, and results of the measurement and analysis process with higher-level management and resolve issues.	Most KPAs Ve 1	Not directly addressed
	Process and Product Quality Assurance	SG1	Adherence of the performed process and associated work products and services to applicable process descriptions, standards, and procedures is objectively evaluated.	SQA Goal 2	CMM® v1.1 was often interpreted to call for independent SQA who reported directly to the senior management.

Process Maturity Level 2	CMMI® Process Area	CMMI® Goal	CMMI® Specific Practice or Generic Practice	SW-CMM® V1.1 Goal/Common Feature	Comments
					In CMMI® v1.1, the introductory material for PPQA points out that organizations can have other safeguards besides independence to ensure objectivity.
			SP 1.1. Objectively evaluate the designated performed processes against the applicable process descriptions, standards, and procedures.	SPE Me 2, Ve 3 SQA Ac 4	
			SP 1.2. Objectively evaluate the designated work products and services against the applicable process descriptions, standards, and procedures.	SPE Me 1, Ve 3 SQA Ac 5	
		SG2	Noncompliance issues are objectively tracked and communicated, and resolution is ensured.	SQA Goal 4	
			SP 2.1. Communicate quality issues and ensure resolution of noncompliance issues with the staff and managers.	SQA Ac 6, 7	
			SP 2.2. Establish and maintain records of the quality assurance activities.	SQA Ac 4, 5, 7	Not directly addressed
		GG 2	The process is institutionalized as a managed process.	Implied by Level 2	
			GP 2.1. Establish and maintain an organizational policy for planning and performing the process and product quality assurance process.	SQA Co 1	

Process Maturity Level 2	CMMI® Process Area	CMMI® Goal	CMMI® Specific Practice or Generic Practice	SW-CMM® V1.1 Goal/Common Feature	Comments
			GP 2.2. Establish and maintain the plan for performing the process and product quality assurance process.	SQA Ac 1 SQM Ac 1,2	SW-CMM® v1.1 doesn't always specify "maintain"
			GP 2.3. Provide adequate resources for performing the process and product quality assurance process, developing the work products, and providing the services of the process.	SQA Ab 2	
			GP 2.4. Assign responsibility and authority for performing the process, developing the work products, and providing the services of the process and product quality assurance process	SQA Ab 1	
			GP 2.5. Train the people performing or supporting the process and product quality assurance process as needed.	SQA Ab 3,4	
			GP 2.6. Place designated work products of the process and product quality assurance process under appropriate levels of configuration management.	SCM Goal 2	
			GP 2.7. Identify and involve the relevant stakeholders of the process and product quality assurance process as planned.	SCM Ac 9 SQA Ac 1	"Relevant stakeholder" = "affected groups"
			GP 2.8. Monitor and control the process and product quality assurance process against the plan for performing the process and take appropriate corrective action.	SQA Me 1	

Process Maturity Level 2	CMMI® Process Area	CMMI® Goal	CMMI® Specific Practice or Generic Practice	SW-CMM® V1.1 Goal/Common Feature	Comments
			GP 2.9. Objectively evaluate adherence of the process and product quality assurance process against its process description, standards and procedures, and address noncompliance.	SQA Ve 3	
			GP 2.10 Review the activities, status, and results of the process and product quality assurance process with higher-level management and resolve issues.	SQA Ve 1	
	Configuration Management	SG1	Baselines of identified work products are established and maintained.	SCM Goal 2	
			SP 1.1. Identify the configuration items, components, and related work products that will be placed under configuration management.	SCM Ac 4	
			SP 1.2. Establish and maintain a configuration management and change management system for controlling work products.	SCM Ac 3,5	
			SP 1.3. Create or release baselines for internal use and for delivery to the customer.	SCM Ac 7	
			SG 2. Changes to the work products under configuration management are tracked and controlled.	SCM Goal 2,3	
			SP 2.1. Track change requests for the configuration items.	SCM Ac 5	
			SP 2.2. Control changes to the configuration items.	SCM Ac 5,6	

Process Maturity Level 2	CMMI® Process Area	CMMI® Goal	CMMI® Specific Practice or Generic Practice	SW-CMM® V1.1 Goal/Common Feature	Comments
			SG 3. Integrity of baselines is established and maintained.	SCM Goal 3	
			SP 3.1. Establish and maintain records describing configuration items.	SCM Ac 4,8	
			SP 3.2. Perform configuration audits to maintain integrity of the configuration baselines.	SCM Ac 10, Ve 3	
		GG2	The process is institutionalized as a managed process.	Implied by Level 2.	
			GP 2.1. Establish and maintain an organizational policy for planning and performing the configuration management process.	SCM Co 1	
			GP 2.2. Establish and maintain the plan for performing the configuration management process.	SCM Ac 1, 2	SW-CMM® v1.1 doesn't specify "maintain"
			GP 2.3. Provide adequate resources for performing the configuration management process, developing the work products, and providing the services of the process.	SCM Ab 3	
			GP 2.4. Assign responsibility and authority for performing the process, developing the work products, and providing the services of the configuration management process.	SCM Ab 1, 2	
			GP 2.5. Train the people performing or supporting the configuration management process as needed.	SCM Ab 4, 5	

Process Maturity Level 2	CMMI® Process Area	CMMI® Goal	CMMI® Specific Practice or Generic Practice	SW-CMM® V1.1 Goal/Common Feature	Comments
			GP 2.6. Place designated work products of the configuration management process under appropriate levels of configuration management.	SCM Goal 2	
			GP 2.7. Identify and involve the relevant stakeholders of the configuration management process as planned.	SCM Ac 1,2,9	
			GP 2.8. Monitor and control the configuration management process against the plan for performing the process and take appropriate corrective action.	SCM Me 1	
			GP 2.9. Objectively evaluate adherence of the configuration management process against its process description, standards, and procedures, and address noncompliance.	SCM Ve 4	
			GP 2.10. Review the activities, status, and results of the configuration management process with higher-level management and resolve issues.	SCM Ve 1	

References

IEEE PUBLICATIONS

1. IEEE/ANSI. IEEE Guide to software configuration management. ANSI/IEEE Std 1042-1987, IEEE Press, New York, NY, USA. 1987.
2. IEEE Standard Glossary of Software Engineering Terminology, IEEE Std 610.12-1990 (Sept 28) Reaffirmed Sept 2002, IEEE Press, New York, NY, USA. 2002.
3. IEEE Standard for Software Quality Assurance Plans, IEEE Std 730-2002 (Sept), IEEE Press, New York, NY, USA. 2002.
4. IEEE Standard for Software Configuration Management Plans, IEEE Std 828-1998 (Jun 25), IEEE Press, New York, NY, USA. 1998.
5. IEEE Standard for Software Test Documentation, IEEE Std 829-1998 (Sep 16), IEEE Press, New York, NY, USA. 1998.
6. IEEE Recommended Practice for Software Requirements Specifications, IEEE Std 830-1998 (Jun 25), IEEE Press, New York, NY, USA. 1998.
7. IEEE Standard Dictionary of Measures to Produce Reliable Software, IEEE Std 982.1-1988 (Jun 9), IEEE Press, New York, NY, USA. 1988.
8. An American National Standard—IEEE Standard for Software Unit Testing, ANSI/IEEE Std 1008-1987(R1993), Reaffirmed Dec 2002, IEEE Press, New York, NY, USA. 2002.
9. IEEE Standard for Software Verification and Validation, IEEE Std 1012-1998 (Mar 9), IEEE Press, New York, NY, USA. 1998.
10. Supplement to IEEE Standard for Software Verification and Validation: Content Map to IEEE/EIA 12207.1-1996, IEEE Std 1012a-1998 (Sept 16), IEEE Press, New York, NY, USA. 1998.
11. IEEE Recommended Practice for Software Design Descriptions, IEEE Std 1016-1998 (Sep 23), IEEE Press, New York, NY, USA. 1998.
12. IEEE Standard for Software Reviews, IEEE Std 1028-1997 (Mar 4) Reaffirmed Sept 2002, IEEE Press, New York, NY, USA. 2002.

13. IEEE Standard Classification for Software Anomalies, IEEE Std 1044-1993 (Dec 2) Reaffirmed Sept 2002, IEEE Press, New York, NY, USA. 2002.
14. IEEE Standard for Software Productivity Metrics, IEEE Std 1045-1992 (Sept 17) Reaffirmed Dec 2002, IEEE Press, New York, NY, USA. 2002.
15. IEEE Standard for Software Project Management Plans, IEEE Std 1058-1998 (Dec 8), IEEE Press, New York, NY, USA. 1998.
16. IEEE Standard for a Software Quality Metrics Methodology, IEEE Std 1061-1998 (Dec 8), IEEE Press, New York, NY, USA. 1998.
17. IEEE Recommended Practice for Software Acquisition, IEEE Std 1062-1998 Edition (Dec 2) Reaffirmed Sept 2002, IEEE Press, New York, NY, USA. 2002.
18. IEEE Standard for Software User Documentation, IEEE Std 1063-2001 (Dec 5), IEEE Press, New York, NY, USA. 2001.
19. IEEE Standard for Developing Software Life Cycle Processes, IEEE Std 1074-1997 (Dec 9), IEEE Press, New York, NY, USA. 1997.
20. IEEE Standard Reference Model for Computing System Tool Interconnections, IEEE Std 1175-1991 (Dec 5), IEEE Press, New York, NY, USA. 1991.
21. IEEE Guide for CASoftware Engineering Tool Interconnections—Classification and Description, IEEE Std 1175.1-2002 (Nov. 11), IEEE Press, New York, NY, USA. 2002.
22. IEEE Standard for Software Maintenance, IEEE Std 1219-1998 (Jun 25), IEEE Press, New York, NY, USA. 1998.
23. IEEE Standard for the Application and Management of the Systems Engineering Process, IEEE Std 1220-1998 (Dec 8), IEEE Press, New York, NY, USA. 1998.
24. IEEE Standard for Software Safety Plans, IEEE Std 1228-1994 (Mar 17) Reaffirmed Dec. 2002, IEEE Press, New York, NY, USA. 2002.
25. IEEE Guide for Developing System Requirements Specifications, IEEE Std 1233, 1998 Edition (Apr 17), Reaffirmed Sept 2002, IEEE Press, New York, NY, USA. 2002.
26. IEEE Standard for Functional Modeling Language—Syntax and Semantics for IDEF0, IEEE Std 1320.1-1998 (Jun 25), IEEE Press, New York, NY, USA. 1998.
27. IEEE Standard for Conceptual Modeling Language Syntax and Semantics for IDEF1X 97 (IDEF object), IEEE Std 1320.2-1998 (Jun 25), IEEE Press, New York, NY, USA. 1998.
28. IEEE Guide for Information Technology—System Definition—Concept of Operations (ConOps) Document, IEEE Std 1362-1998 (Mar 19), IEEE Press, New York, NY, USA. 1998. IEEE Standard for Information Technology—Software Reuse—Data Model for Reuse Library Interoperability: Basic Interoperability Data Model (BIDM), IEEE Std 1420.1-1995 (Dec 12), Reaffirmed June 2002, IEEE Press, New York, NY, USA. 2002.
29. Supplement to IEEE Standard for Information Technology—Software Reuse—Data Model for Reuse Library Interoperability: Asset Certification Framework, IEEE Std 1420.1a-1996 (Dec 10), Reaffirmed June 2002, IEEE Press, New York, NY, USA. 2002.
30. IEEE Trial-Use Supplement to IEEE Standard for Information Technology—Software Reuse—Data Model for Reuse Library Interoperability: Intellectual Property Rights Framework, IEEE Std 1420.1b-1999 (Jun 26), Reaffirmed June 2002, IEEE Press, New York, NY, USA. 2002.
31. IEEE Standard—Adoption of International Standard ISO/IEC 14102: 1995—Information Technology—Guideline for the Evaluation and Selection of CASoftware Engineering Tools, IEEE Std 1462-1998 (Mar 19), IEEE Press, New York, NY, USA. 1998.
32. IEEE Standard—Adoption of International Standard ISO/IEC 12119: 1994(E)—Information Technology—Software packages—Quality requirements and testing, IEEE Std 1465-1998 (Jun 25), IEEE Press, New York, NY, USA. 1998.

33. IEEE Recommended Practice for Architectural Description of Software Intensive Systems, IEEE Std 1471-2000 (Sept 21), IEEE Press, New York, NY, USA. 2000.
34. IEEE Guide—Adoption of PMI Standard—A Guide to the Project Management Body of Knowledge, IEEE Std 1490-2003 (Dec 10) Replaces 1490-1998 (Jun 25), IEEE Press, New York, NY, USA. 2003.
35. EIA/IEEE Interim Standard for Information Technology—Software Life Cycle Processes—Software Development: Acquirer-Supplier Agreement, IEEE Std 1498-1995 (Sept 21), IEEE Press, New York, NY, USA. 1995.
36. IEEE Standard for Information Technology—Software Life Cycle Processes—Reuse Processes, IEEE Std 1517-1999 (Jun 26)
37. IEEE Standard for Software Life Cycle Processes—Risk Management, IEEE Std 1540-2001 (Mar 17), IEEE Press, New York, NY, USA. 2001.
38. IEEE Recommended Practice for Internet Practices—Web Page Engineering—Intranet/Extranet Applications IEEE Std 2001-2002 (Jan 21, 2003), IEEE Press, New York, NY, USA. 2003.
39. Industry Implementation of International Standard ISO/IEC 12207:1995, Standard for Information Technology—Software Life Cycle Processes, IEEE/EIA 12207.0/.1/.2-1996 (Mar), IEEE Press, New York, NY, USA. 1996.
40. IEEE, IEEE Standards Collection, Software Engineering, 1994 Edition, IEEE Press, New York, NY, USA, 1994.
41. IEEE, IEEE Software Engineering Standards Collection, IEEE Press, New York, NY, USA, 2003.
42. IEEE, Software and Systems Engineering Standards Committee Charter Statement, http://standards.computer.org/S2ESC/S2ESC_pols/S2ESC_Charter.htm, USA, 2003.
43. S2ESC Guide for Working Groups, http://standards.computer.org/S2ESC/S2ESC_wgresources/S2ESC-WG-Guide-2003-07-14.doc, USA, 2003.
44. IEEE, Guide to the Software Engineering Body of Knowledge (SWEBOK), Trial Version, IEEE Press, New York, NY, USA, 2001.
45. McConnell, S., "The Art, Science, and Engineering of Software Development," *IEEE Software Best Practices, 15,* 1, 1998.

SEI PUBLICATIONS

46. Bothwell, C. and Masters, S., "CMM® Appraisal Framework Version 1.0," Software Engineering Institute, Carnegie Mellon University, Technical Report, CMU/SEI-95-TR-001, Carnegie Mellon University, Pittsburgh, PA, USA, 1995.
47. Bounds, N.M. and Dart S.A., *Configuration Management (CM) Plans: The Beginning to your CM Solution,* Software Engineering Institute, Carnegie Mellon University, Pittsburgh, PA, USA, 1993.
48. *The Capability Maturity Model: Guidelines for Improving the Software Process,* v.1.1, Software Engineering Institute, Carnegie Mellon University, Pittsburgh, PA, USA, 1997.
49. Dunaway, D.K., *CMM®* Based Appraisal for Internal Process Improvement (CBA IPI) Team Member's Handbook, v1.1, Software Engineering Institute, Carnegie Mellon University, Handbook, CMU/SEI-96-HB-005, Carnegie Mellon University, Pittsburgh, PA, USA, 1996.
50. Dunaway, D.K. and Masters S., "CMM-Based Appraisal for Internal Process Improvement (CBA IPI) Method Description," Software Engineering Institute, Carnegie Mellon University, Technical Report, CMU/SEI-96-TR-007, Carnegie Mellon University, Pittsburgh, PA, USA, 1996.

/ Gibbs, N., "A Mature Profession of Software Engineering," Software Engineering arnegie Mellon University, Technical Report, CMU/SEI-96-TR-004, Carnegie Mellon ／sity, Pittsburgh, PA, USA, 1996.

J. and Myers, C., "The IDEAL Model: A Practical Guide for Improvement," *Bridge*, Software Engineering Institute, Carnegie Mellon University, Pittsburgh, PA, 1997.

s, S. and Bothwekk, C., "CMM® Appraisal Framework, Version 1.0," Software Engi-ɩg Institute, Carnegie Mellon University, Technical Report, CMU/SEI-95-TR-001, ADA 2ყ⊃⌐00, Carnegie Mellon University, Pittsburgh, PA, USA, 1995.

54. McFeeley, B., *IDEAL: A User's Guide to Software Process Improvement,* Software Engineering Institute, Carnegie Mellon University, Handbook, CMU/SEI-96-HB-001, Carnegie Mellon University, Pittsburgh, PA, USA, 1996.
55. Paulk, M. C., Weber, C. V., Garcia, S. M., Chrissis, M. B., and Bush, M., *Key Practices of the Capability Maturity Model, Version 1.1,* Software Engineering Institute, Carnegie Mellon University, Pittsburgh, PA, USA, 1993.
56. SEI, *Appraisal Requirements for CMMI®* Version 1.1 (ARC, V1.1), Software Engineering Institute, Carnegie Mellon University, Technical Report, CMU/SEI-2001-TR-034, Carnegie Mellon University, Pittsburgh, PA, USA, 2001.
57. SEI, *Capability Maturity Model Integration (CMMI) for Software Engineering,* v.1.1 Staged Representation, Software Engineering Institute, Carnegie Mellon University, Technical Report, CMU/SEI-2002-TR-029, Carnegie Mellon University, Pittsburgh, PA, USA, 2002.
58. SEI, *CMM®* Based Appraisal: Team Training Participant's Guide, CMM® Based Appraisal for Internal Process Improvement (CBA IPI), Software Engineering Institute, Carnegie Mellon University, Pittsburgh, PA, USA, 1998.
59. SEI, *Standard CMMI®* Appraisal Method for Process Improvement (SCAMPI), V1.1: Method Definition Document, Software Engineering Institute, Carnegie Mellon University, Pittsburgh, PA, USA, 2001.
60. Zubrow, D., Hayes, W., Siegel, J., and Goldenson, D., *Maturity Questionnaire,* Special Report, CMU/SEI-94-SR-7, Software Engineering Institute, Carnegie Mellon University, Pittsburgh, PA, USA, 1994.

OTHER REFERENCES

61. Arthur, L., *Software Evolution: The Software Maintenance Challenge,* Wiley, 1988.
62. Babich, W., *Software Configuration Management,* Addison-Wesley, 1986.
63. Bersoff, E., Henderson, V., and Siegel, S., *Software Configuration Management: A Tutorial,* IEEE Computer Society Press, 1980, pp. 24–32.
64. Croll, P., "Eight Steps to Success in CMMI-Compliant Process Engineering, Strategies and Supporting Technology," presented at the Third Annual CMMI® Technology Conference and Users Group, USA, 2003.
65. Croll, P., "How to Use Standards as Best Practice Information Aids for CMMI-Compliant Process Engineering," presented at the 14th Annual DoD Software Technology Conference, USA, 2002.
66. Dunn, R.H. and Ullman, R.S., *TQM for Computer Software,* 2nd Edition, McGraw-Hill, 1994.
67. Dymond, K.M., *A Guide to the CMM, Understanding the Capability Maturity Model for Software,* Process Transition International, 1998.
68. Freedman, D.P. and Weinberg, G.M., *Handbook of Walkthroughs, Inspections, and Technical Reviews, Evaluating Programs, Projects, and Products,* 3rd Edition, Dorset House, 1990.
69. Glazer, H., "Two Key Challenges for Implementing CMM," http://www.entinex.com/ImplementingCMMI_page1.cfm, USA, 2004.

70. Hass, A.M.J., *Configuration Management Principles and Practice,* Addison-Wesley, 2003.
71. Hefner, R., "New CMMI® Requirements for Risk Management," *CrossTalk The Journal of Defense Software Engineering,* Feb 2000.
72. Jalote P., *An Integrated Approach to Software Engineering,* 2nd Edition, Springer-Verlag, 1997.
73. Land, S.K., "1st User's of Software Engineering Standards Survey," IEEE Software and Systems Engineering Standards Committee (S2ESC), USA, 1997.
74. Land S.K., "2nd User's of Software Engineering Standards Survey," IEEE Software and Systems Engineering Standards Committee (S2ESC), USA, 1999.
75. Land, S.K., "IEEE Standards User's Survey Results," in *ISESS '97 Conference Proceedings,* IEEE Press, 1997.
76. Land, S.K., "Second IEEE Standards User's Survey Results," in *ISESS '99 Conference Proceedings,* IEEE Press, 1999.
77. McConnell, S., *Professional Software Development,* Addison-Wesley, 2004.
78. Moore, J., "Increasing the Functionality of Metrics through Standardization," presented at the Conference on Developing Strategic I/T Metrics, USA, 1998.
79. Pressman, R., *Software Engineering,* McGraw-Hill, 1987.
80. Royce, Walker, "CMM® vs. CMMI, From Conventional to Modern Software Management," *The Rational Edge,* 2002.
81. Schach, S.R., *Classical and Object-Oriented Software Engineering,* 3rd Edition, Irwin Press, 1993.
82. USAF Software Technology Support Center (STSC), *CMM-SE/SW V1.1to SW-CMM® V1.1 Mapping,* 2002.
83. Veenendall, E.V., Ammerlaan, R., Hendriks, R., van Gensewinkel, V., Swinkels, R., and van der Zwan, M., "Dutch Encouragement; Test standards We Use in Our Projects," *Professional Tester,* Number 16, October 2003.
84. Westfall, L.L., "Seven Steps to Designing a Software Metric," BenchmarkQA, Whitepaper, USA, 2002.
85. Whitgift, D., *Methods and Tools for Software Configuration Management,* Wiley, 1991.
86. Wiegers, K.E., *Creating A Software Engineering Culture,* Dorset House Publishing, 1996.

Index

ARC, 24

Capability maturity model-based assessment of internal process improvement, 12
Change enhancement requests, 42
CMM, 1, 2, 3, 7, 8, 10, 11, 12, 15, 16, 20, 27, 28, 29, 30, 31, 32, 33, 34, 35, 36, 38, 39, 40, 41, 42, 44, 45, 46, 47, 48, 49, 50, 51, 52, 53, 58, 60, 61, 62, 63, 64, 65, 66, 67, 72, 73, 74, 75, 76, 77, 78, 79, 81, 84, 85, 88, 89, 90, 92, 93, 94, 95, 97, 98, 99, 100, 101, 102, 105, 106, 107, 117, 124, 125, 126, 127, 128, 129, 130, 131, 132, 145
   history, 27
   versus CMMI, 27
CMM® AC1, 106
CMM® AC10, 58
CMM® AC4, 93
CMM® AC5, 81
CMM® AC8, 67
CMMI, 1, 2, 3, 4, 15, 16, 17, 18, 19, 20, 21, 22, 23, 24, 27, 28, 29, 30, 31, 32, 33, 34, 38, 39, 40, 41, 42, 44, 45, 46, 47, 48, 53, 54, 55, 56, 57, 58, 60, 68, 69, 70, 71, 72, 74, 79, 80, 81, 83, 84, 85, 86, 90, 91, 92, 94, 95, 96, 102, 103, 104, 105, 106, 107, 108, 111, 112, 113, 114, 115, 124, 125, 127, 128, 129, 130, 132, 145, 146, 147, 148, 149, 150, 151, 152, 153, 154, 155, 156, 157, 158, 159, 160, 161, 162, 163
CMMI®-SW (Staged) Goals for Configuration Management, 95
CMMI®-SW (Staged) Goals for Project Planning, 60
CMMI®-SW (Staged) Goals for Supplier Agreement Management, 107
CMMI®-SW (Staged) Goals Requirements Management., 46
CMMI®-SW (Staged) SP 1.1, 81, 106
CMMI®-SW (Staged) SP 1.2, 114
CMMI®-SW (Staged) SP 1.7, 72
CMMI®-SW (Staged) SP1.1, 94
CMMI®-SW (Staged) SP1.4, 58
CMMI®-SW (Staged) Supplier Agreement Management Analysis, 105
CMMI-SW, 15, 27, 32, 48, 96, 132
   appraisal, 22
   common features, 20
   components, 19
   configuration management, 22
   continuous versus staged, 15
   engineering, 17
   key process areas, 16
   measurement and analysis, 22

CMMI-SW *(continued)*
  process and product quality assurance, 21
  process areas, 21
  process management, 17
  project management, 17
  project monitoring and control, 21
  project planning, 21
  requirements management, 21
  specific and generic goals, 18
  specific and generic practices, 18
  structural elements of, 16
  supplier agreement management, 22
  support, 18
Commitment to perform, 8, 19, 20
Commitments, 30, 34, 47, 48, 49, 51, 55, 58, 60, 61, 65, 67, 69, 72, 95, 96, 98, 99, 107, 117, 126, 128, 129, 149
Configuration management, 30
Contracts, 3, 13, 24, 96, 97, 98, 99, 101, 102, 103, 104, 105, 106, 108
Contractors, 27, 96, 98, 99, 100, 101, 105, 107, 126
Costs, 49, 58, 63, 65, 112, 129
Critical computer resources, 49, 63, 67

Data, 21, 29, 31, 32, 42, 49, 51, 54, 58, 64, 65, 69, 73, 94, 95, 108, 109, 110, 111, 112, 113, 114, 123, 129, 136, 138, 140, 141, 148, 151, 156, 157
  collection, 31, 111, 112, 114
Databases, 42, 87, 94, 129, 139, 141
Differences between CMM® and CMMI-SW® (Staged), 27
  configuration management, 30
  maturity level comparison, 28
  measurement and analysis, 30
  monitoring, 30
  project management, 30
  project planning, 29
  quality assurance, 30
  requirements management, 29
Documented procedure, 49, 50, 51, 58, 64, 75, 77, 88, 89, 98, 99, 100, 101, 106, 126

Estimates, 21, 22, 33, 47, 48, 49, 53, 58, 60, 61, 63, 67, 68, 70, 107, 121, 126, 148
Estimating, 22, 52, 60

Framework, 1, 2, 3, 4, 12, 35, 48, 62, 68, 109, 115, 128, 135, 136, 137, 141
Funding, 37, 40, 41, 44, 45, 52, 66, 77, 82, 89, 101

Goals

  for CMM® Requirements Management, 33
  for CMM® Software Project Planning, 47
  for CMM® Software Quality Assurance, 73
  for CMMI®-SW (Staged) Configuration Management, 85
  for CMMI®-SW (Staged) Process and Product Quality Assurance, 74
  for CMMI®-SW (Staged) Project Planning, 47
  for CMMI®-SW (Staged) Requirements Management, 34
  for CMMI®-SW (Staged) Supplier Agreement Management, 96
  for Software Configuration Management, 94
  of CMMI® Measurement and Analysis, 108
  of Software Project Tracking and Oversight, 61

IDEAL, 12, 28, 117, 118
IEEE and CMM® Software Project Tracking and Oversight, 61
IEEE and CMMI®-SW (Staged) Software Project Monitoring and Control, 68
IEEE Guide for Software Quality Assurance Planning IEEE Std 730.1-1998., 75
IEEE KPA Support
  for Measurement and Analysis, 114
  for Project Monitoring and Control, 72
  for Project Tracking and Oversight, 67
  for Requirements Management, 41
  for Software Configuration Management, 93
  for Software Project Planning, 58
  for Software Quality Assurance, 81
  for Software Subcontractor Management, 106
IEEE Recommended Practice for Software Acquisition IEEE Std 1062-1998, 97
IEEE Software Engineering Standards, 3
  categories of, 4
  development of, 5
  motivation for, 3
IEEE Software Project Management Plan IEEE Std 1058, 48
IEEE Software Requirements Specification IEEE Std 830, 34
IEEE Standard Classification for Software Anomalies IEEE Std 1044™-1993 (R2002), 109
IEEE Standard for Developing Software Life

Cycle Processes IEEE Std 1074™-1997, 109
IEEE Standard for Software Configuration Management Plans IEEE Std 828-1998, 86
IEEE Standard for Software Maintenance IEEE Std 1219, 58
IEEE Standard for Software Productivity Metrics IEEE Std 1045-1992 (R2003), 109
IEEE Standard for Software Quality Assurance Plans IEEE Std 730-2002., 75
IEEE Standard for Software Quality Metrics Methodology IEEE Std 1061™-1998, 62
IEEE Standard for Software Reviews IEEE Std 1028, 68
IEEE Standard for Software Test Documentation IEEE Std 829, 58
IEEE Standard IEEE Standard Dictionary of Measures to Produce Reliable Software IEEE Std 982.1, 109
IEEE standards, 1, 2, 3, 4, 5, 7, 32, 33, 35, 41, 45, 46, 48, 52, 54, 60, 62, 67, 72, 75, 83, 87, 97, 113, 115, 117, 122, 126, 130, 131, 132, 133, 142
IEEE Std 1058 Section 4.5 Managerial Process Plans, 58
IEEE Std 1058 Section 4.5.3.2 Schedule Control Plan, 67, 72
IEEE Std 1062 Section 5.3 Defining the Software Requirements, 106
IEEE Std 1062, Section 4.2, Identify Software Quality Metrics, 114
IEEE Std 1062, Section 4.3.1, Define the data collection procedures., 114
IEEE Std 730 Section 4.6 Software Reviews, 81
IEEE Std 828 Section 4.3.1, Configuration Identification, 94
IEEE Std 830—4.4 Joint Preparation of the SRS, 42
IEEE System Requirements Specification IEEE Std 1233, 35
IEEE-supported process improvement, 117
 define and train the process team (initiate), 117
 software engineering, 117
Implementation pitfalls, 129
 being overly prescriptive, 129
 documentation, 130
 lack of incentives, 130
 no metrics, 130
 remaining confined to a specific stage, 129

Level 2 key process areas, 9
Life cycle process, 50, 107, 109, 135, 136

Managing the software project, 10, 33, 47, 66
Maturity level, 7, 8, 9, 10, 11, 15, 16, 17, 18, 22, 23, 24, 25, 28, 31, 129, 130, 145
Measurement and analysis, 30, 108
Milestones, 30, 47, 48, 61, 64, 67
Milestone review, 69, 72

Process and product quality assurance, 30
Process improvement, 1, 12, 15, 16, 19, 22, 23, 24, 25, 27, 28, 30, 31, 59, 85, 117, 119, 124, 125, 126, 127, 128, 129, 130, 131, 132
Project manager, 2, 52, 74, 78, 121, 127
Project monitoring and control, 30, 68
 goals of, 68
Project monitoring and control analysis, 72
Project planning, 29
Project proposal team, 51, 54, 56
Project software manager, 51, 65, 66

Requirements management, 10, 29
 analysis, 41
Requirements traceability, 42
Resources, 8, 20, 21, 23, 29, 30, 36, 37, 39, 40, 41, 44, 45, 46, 47, 52, 54, 55, 56, 58, 60, 63, 65, 66, 67, 71, 77, 79, 80, 82, 84, 86, 87, 89, 91, 92, 101, 103, 112, 130, 147, 148, 149, 150, 152, 155, 157, 160, 162
Risk, 11, 17, 23, 24, 30, 32, 42, 47, 50, 52, 54, 55, 59, 60, 61, 64, 67, 69, 72, 82, 105, 124, 125, 130, 132, 134, 135, 148, 151
 management, 17, 24, 32, 50, 52, 54, 55, 59, 60, 64, 69, 82, 105, 124, 125, 135

SCAMPI, 22, 24, 25, 128
Self-Audit Using CMMI®-SW (Staged) KPAs, 128
Self-audit using SW-CMM® KPAs, 128
Senior management, 11, 52, 73, 77, 78, 81, 83, 90, 117, 126, 127, 130
Software acquisition, 13, 27, 31, 95, 97, 101, 104, 105, 106, 107, 108, 124, 132, 133
Software anomalies, 109, 113, 115, 122, 140
Software capability evaluation, 13

Software configuration management, 9, 10, 11, 12, 82, 83, 84, 85, 86, 87, 88, 89, 90, 92, 93, 94, 95, 114, 115, 120, 123, 124, 126, 127, 131, 132, 137, 138, 145
Software development, 1, 2, 4, 27, 33, 42, 46, 48, 60, 67, 75, 81, 84, 107, 119, 122, 123, 127, 129
  plan, 47, 52, 61, 66, 67, 98, 105, 107, 127
Software effort, 22, 33, 34, 47, 60, 61, 63, 65, 84, 113, 114
Software engineering group, 29, 33, 35, 36, 41, 46, 51, 54, 64
Software life cycle, 11, 50, 62, 68, 74, 85, 109, 135, 137, 140, 142
Software maintenance, 37, 50, 51, 54, 56, 58, 62, 64, 67, 69, 120, 123, 139
Software measurement, 67, 108, 115
Software planning data, 49
Software productivity metrics, 60, 109, 115
Software project, 2, 10, 11, 12, 21, 22, 33, 34, 35, 41, 46, 47, 48, 49, 50, 51, 52, 55, 56, 58, 60, 61, 62, 64, 66, 68, 69, 73, 74, 75, 77, 78, 81, 82, 83, 87, 88, 94, 105, 107, 109, 113, 114, 119, 126, 129, 130, 132, 138, 139, 145
  planning, 9, 10, 11, 12, 46, 47, 48, 49, 50, 51, 52, 53, 58, 60, 107, 125, 126, 131, 145
  tracking and oversight, 9, 10, 11, 12, 60, 61, 62, 63, 64, 65, 66, 67, 126, 131, 145
  tracking and oversight analysis, 67
Software quality assurance, 9, 10, 11, 12, 38, 53, 66, 73, 74, 75, 76, 77, 78, 81, 83, 90, 94, 100, 102, 114, 115, 126, 127, 131, 132, 137, 145
Software quality metrics, 62, 64, 65, 66, 69, 102, 114, 115, 132
Software requirements, 21, 29, 33, 34, 35, 36, 37, 39, 41, 43, 45, 46, 59, 82, 105, 106, 108, 120, 122, 123, 125, 126, 127, 132, 139
Software review, 35, 68, 79, 80, 81, 83, 122
Software schedule, 49, 65, 67
Software subcontract management, 9, 10, 12, 95, 96, 131, 145
Software subcontract/supplier management, 95
  goals for, 95

Software subcontractors, 95, 96, 97, 98, 99, 100, 101, 102, 105, 106, 107, 126
Software test documentation, 58
Software work products, 47, 49, 50, 61, 63, 66, 73, 74, 75, 81, 82, 84, 85, 88, 93, 94, 126
Structural elements of the SW-CMM® V 1.1 SW-CMM® Maturity Levels, 7
Subcontracted work, 96
Subcontractors, 12, 59, 87, 93, 94, 95, 96, 98, 99, 100, 101, 102, 103, 104, 105, 106, 107, 114, 126
Supporting IEEE software engineering standards, 34, 48, 86
SW-CMM, 1, 7, 9, 10, 15, 22, 27, 28, 29, 30, 31, 32, 45, 46, 47, 48, 60, 74, 81, 83, 94, 95, 105, 106, 107, 125, 126, 128, 132, 145, 146, 147, 148, 149, 150, 151, 152, 153, 154, 155, 156, 157, 158, 159, 160, 161, 162, 163
SW-CMM® Goals for Software Project Planning, 60
SW-CMM® Goals for Software Quality Assurance, 81
SW-CMM® Goals for Software Requirements Management, 45
SW-CMM® Goals for Software Subcontractor Management, 107
SW-CMM® Software Subcontractor Management Analysis, 105
SWEBOK, 119
System requirements specification, 35

Timelines, 125
Tracking, 8, 9, 10, 11, 12, 22, 30, 42, 47, 49, 50, 60, 61, 62, 63, 64, 65, 66, 67, 73, 81, 84, 96, 98, 99, 102, 106, 126, 129, 131, 145
Training, 2, 8, 9, 10, 11, 17, 22, 31, 32, 37, 40, 41, 44, 45, 52, 54, 56, 59, 60, 66, 71, 77, 78, 80, 82, 86, 89, 92, 93, 101, 104, 112, 113, 114, 117, 118, 119, 122, 125, 126, 135, 138, 145

Work product, 11, 18, 20, 21, 22, 30, 34, 36, 38, 39, 40, 46, 47, 49, 50, 53, 56, 58, 60, 61, 63, 66, 67, 68, 71, 72, 73, 74, 75, 77, 79

ABOUT THE AUTHOR

Susan K. Land is the Software Engineering Section Manager for Northrop Grumman/TASC in Huntsville, Alabama. She has participated in many software process improvement efforts, CMM® CAF-based appraisals, and CBA IPI self-assessments. Ms. Land is currently serving as a member of the IEEE Standards Advisory Board SAB, the IEEE Software and Systems Engineering Standards Committee (S2ESC) Management Board, the Editorial Board for the IEEE Software Engineering Online publication, and the IEEE CS Professional Practices Committee. She is a participant in the development of the Strawman and Trial versions of the Software Engineering Body of Knowledge (SWEBOK), and is an IEEE Certified Software Development Professional (CSDP).